Islamic
Architecture in India
Second Edition

**Islamic
Architecture in India**
Second Edition

ISBN: 978-81-239-2320-8 (HB)
ISBN: 978-81-239-0783-3 (PB)

Second Edition   2002
   Reprint   2006, 2008, 2010, 2012, 2013, 2014, 2015, 2016, 2017, 2022
First Edition   1996

Published by Satish Kumar Jain and produced by Varun Jain for

CBS Publishers & Distributors Pvt Ltd
4819/XI Prahlad Street, 24 Ansari Road, Daryaganj, New Delhi 110 002, India
Ph: 011-23289259, 23266861, 23266867   Website: www.cbspd.com
Fax: 011-23243014   e-mail: delhi@cbspd.com;
                                                   cbspubs@airtelmail.in.

Corporate Office: 204 FIE, Industrial Area, Patparganj, Delhi 110 092, India
Ph: 011-4934 4934   Fax: 011-4934 4935   e-mail: publishing@cbspd.com;
                                                     publicity@cbspd.com

Branches

- **Bengaluru:** Seema House 2975, 17th Cross, KR Road, Banasankari 2nd Stage, Bengaluru 560 070, Karnataka, India
  Ph: +91-80-26771678/79   Fax: +91-80-26771680   e-mail: bangalore@cbspd.com
- **Chennai:** 7, Subbaraya Street, Shenoy Nagar, Chennai 600 030, Tamil Nadu, India
  Ph: +91-44-26680620, 26681266   Fax: +91-44-42032115   e-mail: chennai@cbspd.com
- **Kochi:** 42/1325, 1326, Power House Road, Opp KSEB, Power House, Ernakulum Kochi 682 018, Kerala, India
  Ph: +91-484-4059061-65,67   Fax: +91-484-4059065   e-mail: kochi@cbspd.com
- **Kolkata:** 147, Hind Ceramics Compound, 1st Floor, Nilgunj Road, Belghoria, Kolkata-700056, West Bengal, India
  Ph: +91-9096713055/7798394118, 9836841399   e-mail: kolkata@cbspd.com
- **Lucknow:** Basement, Khushnuma Complex, 7 Meerabai Marg (Behind Jawahar Bhawan),Lucknow-226001, UP, India
  Ph: +0522-4000032
- **Mumbai:** PWD Shed, Gala no 25/26, Ramchandra Bhatt Marg, Next to JJ Hospital Gate no. 2, Opp. Union Bank of India, Noorbaug, Mumbai-400009, Maharashtra, India
  Ph: 022-66661880/89   e-mail: mumbai@cbspd.com

Representatives

• Hyderabad	0-9885175004	• Jharkhand	0-9811541605	• Nagpur	0-9421945513
• Patna	0-9334159340	• Pune	0-9623451994	• Uttarakhand	0-9716462459

Printed at Neekunj Print Process, Haryana, India

Islamic
Architecture in India
Second Edition

SATISH GROVER

CBS Publishers & Distributors Pvt Ltd

New Delhi • Bengaluru • Chennai • Kochi • Kolkata • Lucknow • Mumbai
Hyderabad • Jharkhand • Nagpur • Patna • Pune • Uttarakhand

Acknowledgement

Having always believed in architecture and language as inviolable matrices of civilisation, it was delightful to discover the parallels in the growth of Urdu and that of Islamic architecture in India. This discovery was the result of innumerable discussions with journalist friend Saeed Naqvi wherein he recited Urdu couplets to prove his point of Hindu-Muslim assimilation and I responded with descriptions of the same process in the growth of Islamic architecture in India.

The sketches for this volume were made by my office staff — Anurag Gupta, B.K. Sharma and student trainees. The photographs have almost entirely been contributed by Rabi Shanker Dey to whom I owe a great debt for his time and patience in bearing with me, as also for his expert help in fine-tuning the design of the book.

The reader will also notice that I have at random quoted phrases from other writers when I have found their descriptions of buildings to be more apt, precise or lyrical than mine. These quotes are all from books listed in bibliography. My greatest indebtedness on this account is to the works of Percy Brown, Hermann Goetz Rawlinson, and *Architecture of the Islamic World* edited by G. Michell.

SATISH GROVER

Preface to the Second Edition

The first edition of this book was published in 1980 by Vikas Publishing House, and that entire print was sold out within two years of the publication.The text was revised and the revised print was brought out by another publisher in 1996, and this printing has also been entirely sold out.

From the time when the book was first printed, the printing technology in India has grown vastly. Also, the healthy economy of the publishing industry has resulted in the publication of books in India of much better standard in all spheres of the book production, for example, printing technology, quality of paper, quality of binding. Also, the books printed and published in India are now being marketed on a world-wide scale. This is particularly applicable to the books of this nature which require very good quality of printing and paper for its illustrations. This fact has encouraged me and the publishers to bring out a new deluxe edition of the book in colour and in a new format which is now possible. I have also taken this opportunity to review and revise the book completely, update the material wherever possible and add fresh illustrations.

I am quite sure that this edition will be welcomed by the audience which comprises intelligent tourists, students of architecture, professionals, and all English speaking libraries of the schools of architecture in the world.

SATISH GROVER
New Delhi, 2001

Preface to the First Edition

This volume covers the impact of resurgent Islamic thought, ideals, religion and philosophy on the ancient and established civilisation of the Hindus in India, a subject that has fascinated me for long. How powerful the force of Islam was and how effectively the Indian builder produced a viable style out of the two seemingly contradictory philosophies is only too apparent in the great and diverse masterpieces of Islamic architecture all over India. The Qutab Minar at Delhi, the Jami and Atala Masjids at Mandu and Jaunpur, the Teen Darwaza at Ahmedabad, the Gol Gumbaz at Bijapur, the city of Fatehpur Sikri and the Taj at Agra are all undoubtedly and distinctly Islamic. This book, however, stresses on the indigenous quality of each of these monuments. In fact, they represent a mirrorlike architectural reflection of the synthesis of Hindu and Muslim cultures that reached its most creative period under the rule of Akbar. This process of assimilation is evident in varying degrees at Delhi, Jaunpur, Bengal, Mandu, Gujarat and the Deccan, in fact all over India that was under the Mughal rule. I trust this emphasis will be evident to the reader, though it may get blurred at times under the ritualistic but necessary descriptions of the innumerable specimens of Islamic architecture in India.

The format of the book is much the same as followed in its companion volume *Buddhist and Hindu Architecture*, which is to be published later. In the context of studying Islamic architecture in India, an overall chronological study of its earliest phase in Delhi is undoubtedly rewarding. Subsequently, in describing the parallel but architecturally distinct styles of the various provinces, it becomes necessary to study each in its own chronological order to understand the evolution of the style within its own geographic, social and artistic parameters. This process is applied to each of the distinct regional styles. When the whole of the subcontinent came under the sway of a single unified empire, it was the writ and personality of each Mughal ruler that strongly influenced the architecture of the times. As such in its final stages, the story is compartmentalised into a study of the achievements of the builders under the guidance of the individual ruling authority.

I hope this book will be of interest to students of architecture, practising architects, and all those who are sensitive to India's rich and diverse cultural heritage.

SATISH GROVER

Contents

Gol Gumbaz,
Bijapur

Introduction

Our story of Islamic architecture opens in the same region where some four thousand years ago the Indus Valley folk had built the earliest known civilization. In the beginning of the seventh century AD, Qasim, a young adventurer of the illustrious Arab tribe of the Quraish, provoked by reports of Hindu piracy of Arab fleets, had captured some of the Hindu cities of the Sind. However, since his audacious spirit was restrained by the advice of moderation sent by the Governor of Baghdad, this invasion of Qasim turned out to be of little political importance — 'an episode in the history of India and Islam, a triumph without results.' Architecturally, too, the earliest footprints of Islam in the Sind Valley in India have been all but obliterated either through natural calamities or by the subsequent looting of the entire sea coast by the Ghaznavids and Ghorids. The meagre remains of the foundations of the earliest Islamic structures on the Indian subcontinent give virtually no information other than that which is merely statistical.

The multiple raids into India by the Turks and Afghans some three hundred and fifty years later than the Arab settlement in the Sind, proved to be of greater political and historical consequence. Though the Ghaznavids and Ghorids too, did not build empires in India, their devastating raids into the land in the early twelfth century laid bare the heroic but feeble defences of India which were manned by the Rajput warriors. Ultimately, the brave army of the Rajputs, led by the legendary figure of Prithviraj Chauhan, could not withstand the repeated onslaughts of the Afghans. In AD 1192 on the historic battlefield of Tarain, 'like a great building, the Hindu host tottered and collapsed in its own ruins.' As we shall see, almost literally out of the ruins grew up a new and refreshing tradition of Islamic building in India.

Within an incredibly short period of this crucial Rajput defeat we find a Muslim Slave Dynasty king established as the ruler in Delhi, who commenced in right earnest to structure the earliest extent mosque on Indian soil — the Quwwat-ul-Islam in the famous Qutb complex south of modern Delhi. In his hurry to carry out Mohammed's dictates of laying out a place of worship for the faithful in the conquered territory, Qutb-ud-din could not wait to import artisans, masons and architects from his native country. And, so, right at the inception of Islamic building activity in India, a sort of joint venture between local Hindu master builders and Muslim overseers was inevitable. In his haste, though Qutb-ud-din did not even have the building material, he was content to remove readymade blocks from existing Hindu and Jaina temples and reorganize them around a rectangular court to quickly assemble the essential rudiments of a mosque. The net result of this earliest effort in mosque architecture in India may well be termed an 'archaeological miscellany.' At the same time, it sowed the seeds of a tradition of give-and-take between the rather austere traditions of Islamic building and the sculptural skills of the local Hindu stone masons. In fact, the subsequent history of Indian Islamic architecture is a fascinating study of the gradual fusion of two seemingly opposite ideals into one of the richest periods of Indian architectural history. For the abundance of building activity during more than 500 years of the Islamic period we must thank the rather vainglorious notions of the Muslim rulers, who were fond of erecting new cities to perpetuate their names in history. Thus, no sooner had Muslim power consolidated itself in Delhi to build impressive victory towers like the Qutb Minar, that ambitious rulers like Ala-ud-din Khalji set out to build complete new cities.

Ala-ud-din laid out his city of Siri in the neighbourhood of the Qutb complex. Though almost nothing of this city has survived, Khalji's more modest Alai Darwaza at the Qutb complex marks the beginning of the process of refinement of the basic module of Islamic architecture — a cubic volume crowned by a hemispherical true dome. Once this basic module had been perfected and the new technological principles of the true arch and the dome understood by the Hindu builders, this module could be assembled together in varying combinations to build mosques, tombs, palaces, markets and entire cities. Armed with this basic vocabulary, the Tughlaqs proceeded to endow their new cities in Delhi with a genuine Islamic urban flavour. This is strongly felt in the fortress cities of Tughlaqabad and Firuz Shah Kotla founded by Ghias-ud-din and Firuz Shah. The latter, laid on the banks of the Jamuna river, enunciated clearly the principles of the planning of a Muslim citadel which consisted of peripheral defensive ramparts and a series of public and private courtyards aligned along a central axis that culminated in the royal private palaces. The Tughlaqs were also responsible for creating a rather militant style with buttresses and circular pylons. Their militant attitude, however, came to naught under the assault of Timur the Mongol, who sacked and laid waste the city just after the end of the rule of Firuz Shah Tughlaq.

This devastation of the Islamic capital at Delhi was a signal for the artisans and craftsmen to migrate to the various regional centres of Muslim power that had sprung up all over India. These were centred around the cities of Jaunpur and Pandua in the east, Ahmedabad in the west, Malwa in the central India and Bijapur in the south. In each of these regions, in the early phase, Muslim buildings were quickly put together from spoils of Hindu temples. Gradually, however, the craftsmen evolved viable and individualistic styles as they began to seriously contend with the climatic, geographic and social circumstances of each region. Thus, in Bengal, for example, the non-availability of stone as a building material and the incessant rains gave rise largely to covered mosques with a characteristic steeply parapet. In Jaunpur, on the other hand, the strong influence of the Tughlaq style of Delhi resulted in the construction of large flamboyant stone pylons interwoven with colonnaded *liwans*. In Mandu, too, the Tughlaqian influence is apparent in the earlier stages. Ultimately, however, the Mandu builders evolved an extremely robust and original style of their own which is seen as its best in the Jami Masjid, which even in its incomplete form is probably one of the finest mosques in India. In Gujarat and its environs, however, the Muslim rulers seem to have been completely swamped by the artistic skills of the local Jaina and Hindu craftsmen. They devised and constructed *liwans* for their mosques which were virtually temple *mandapas* with a facade of pointed arches. So much so that these halls, supported by a virtual forest of columns, have been classified as 'temples in mosques'. In sharp contrast, the South Indian craftsman, early in his career of building for his Muslim overlords, dropped any pretence of adapting Hindu techniques for Islamic buildings. Rather, with immigrants well versed in Persian technology arriving by the sea route on the south-western coast of India, Muslim building activity in this region displays a strong structural bias. Beginning with a large fully covered mosque at Gulbarga, this tradition culminated in the building of the famous Gol Gumbaz at Bijapur, the largest single dome of its time.

In the meanwhile, the Sayyids and Lodis held nominal sway over the cities of Agra and Delhi subsequent to the departure of Timur after his sack of Delhi. During their rule, the dominance of Islamic structural techniques was well established with the construction of massive boat-keel-shaped domes over the numerous tombs of this period. The major contribution of Lodi craftsmen was the adoption of the octagonal tomb. Two of these are to be seen at Delhi which, under the Lodis, had become a royal necropolis. This rather lackadaisical state of affairs in India was given a welcome jolt by yet another Islamic invasion of India, this time that of the Mughals or Mongols, led by the valiant Babur.

Babur's compact and well organized army, spearheaded by a swift cavalry and backed by fire power, defeated the vast but ponderous army of the Lodis in the legendary battlefield of Panipat. Babur, a homeless wanderer for long, had no desire to merely plunder and return. Rather, he was in search of an empire to rule, and the chaotic situation in India was tailor-made for him to assert his authority. Though Babur ruled for a mere four years, he laid the foundations of a kingdom that was to flourish into the famous Mughal empire. This empire, though, was founded as much on Babur's endeavours as on the genius of the Afghan usurper Sher Shah, who forced Babur's son Humayun to relinquish Delhi and flee to Persia. In a brilliant rule of five years Sher Shah set up new systems of administration which were later adopted by Akbar, once the Mughals regained authority. To the architectural tradition, too, Sher Shah contributed handsomely. While the Qila Masjid at Delhi, in the balanced perfection of its facade became a prototype for the Mughals, Sher Shah's own tomb at Sasaram in Bihar is indeed a fitting climax to the series of octagonal tombs erected by the Tughlaqs and Lodis. Sher Shah's traditions, however, were not destined to be kept alive by his dynastic successors. Finally, Humayun managed to oust the Afghans and established the Mughal standard once again at Delhi. His rule was short-lived too. However, as subsequent events proved, Humayun bequeathed a great gift to India — his son Akbar, who took charge of the royal reigns at the tender age of eleven.

With the arrival of Akbar on the Indian scene an era of unparalleled and inspired building activity began. Under his benevolent but powerful guiding hand all that was best in the building tradition of India suddenly came to life. Humayun had brought back from his exile courtiers and craftsmen brimming with Persian ideas. During Akbar's rule these were blended with the Hindu and Buddhist traditions into a style **as unique** as the eclectic personality of Akbar. This is seen at its best in Humayun's tomb at Delhi, in the numerous structures of Akbar's new capital city at Fatehpur Sikri, and in his own tomb at Sikandra. Akbar, during his long tenure, not only vitalized the traditions of Muslim rule in India but bequeathed to his successors an all-India empire the size of which had not existed before. Akbar's son, Jahangir, who had languished as a prince under Akbar's long rule, proved to be none too great a builder of empire of architecture. Nevertheless, he kept alive the traditions of art in his court by bestowing generous patronage on a style of miniature painting. He was also responsible for reviving in a big way the art that his great grandfather Babur had brought to India — that of laying out numerous Mughal gardens — particularly in his favourite valley of Kashmir. Apart from this, it was under this rather indulgent rule that his wife replaced the dignified austerity of Akbar's sandstone architecture with the flamboyance and lustre of pure white Makrana marble.

It was this material that Jahangir's successor, Shahjahan, seized upon with great avidity. The passionate builder that he turned out to be, he took the traditions of Mughal architecture to their climactic best in the famous Taj Mahal at Agra. But before the Taj was built, under Shahjahan's critical eye his craftsmen had mastered the use of marble as a building material. With the growing wealth of the empire the great Mughal now endeavoured to lay out the city of Shahjahanabad at Delhi. The seraglio at the Delhi Fort is studded with exquisite marble pavilions luxuriously embellished and surrounded by gardens and water channels. For the faithful there is the famous Jami Masjid just opposite the Red Fort. This mosque, the Taj Mahal and palaces of Shahjahan were, however, to prove to be the swan song of Islamic architecture in India.

The Mughal empire continued to survive under the orthodox and bigoted rule of Shahjahan's son, Aurangzeb, for another sixty years. However, it is a sad commentary on Aurangzeb that under his patronage, or rather lack of it, not a single work of art of any consequence was created. The traditions of building in India, though, were not quite extinguished. These continued to flourish under the patronage of the Rajput rulers of Rajasthan and central India who employed the immigrant craftsmen from the Mughal courts to build for them cities and palaces. This fallout from the traditions of Mughal architecture and its final extinction under the dry rules and regulations of the British Public Works Department, however, are material for another volume on the history of Indian architecture.

Quwwat-ul-Islam Mosque, Delhi

CHAPTER 1

The Arabs, Afghans and Islamic India

AD 727–AD 1287

In the early eighth century AD Muhammed Qasim, riding on the crest of Arab waves of success from across central Asia to the Spanish fringes of Europe, invaded the Hindu kingdom in the valley of the Sind. The route he had followed across the Makran was the same by which the Sumerians had wandered into India, some four thousand years earlier, to found the Indus Valley civilization. Belonging to the illustrious tribe of Quraish, Muhammed Qasim, however, had a more specific mission to perform. He had been sent by Hajjaj, the famous Muslim governor of Iraq, to protect the sea trade routes of the Arabs with the western coast of India and Ceylon. It would appear that Indian pirates operating from the Sind coast had waylaid a floundering Arab fleet. The audacity of the looters had irked the rising militant Islamic power in Baghdad. The emissary from Baghdad, Muhammed Qasim, did more than merely accomplish the task of restraining the pirates of Debal, conjectured to be the present day Bhambore, situated 40 km east of modern Karachi.

Arab Dictates of Tolerance

Qasim stormed the city of Debal, defeated the Brahmana King Dahir, and pushed right up to the western banks of the Indus. Not content with his success, he went on to capture the cities of Bahmannadad and Alor. Ultimately, even the northern region of Multan capitulated to the Arab invader's zeal. This early incursion of the Arabs into India subsequently turned out to be more of an Islamic cultural delegation than a sustained military adventure. The more destructive and politically oriented invasions were to follow centuries later. In this instance, tolerance in dealing with the Hindus was the directive issued straight from Iraq: 'After they have become Zimmis (protected subjects), we have no right whatever to interfere with their lives or their property. Do, therefore, permit them to build the temples of those they worship,' wrote Hajjaj to the ambitious Muhammed Qasim. This restraint against demolition did not prevent the Arabs from building new mosques in glory of the God they worshipped. This they must have done in plenty. However, in the face of natural and manmade calamities that the region of Sind suffered from subsequently, not a single example of the earlier Arab building efforts is extant today.

Earliest Mosque in India

During recent excavations at the site known today as Bhambore, the remains of a mosque of this period of Arab occupation have been discovered *(Fig 1.01)*. From inscriptions it would seem that the mosque was erected in AD 727, which makes it the earliest known Muslim monument in the Indian subcontinent. Our entire knowledge of this historic monument from excavated ruins is merely statistical. It had a paved courtyard of 75 ft × 58 ft (23 m × 18 m) which was 128 ft × 122 ft (39 m × 37 m) in outer dimensions. On three sides were cloisters with two rows of pillars, and on the west sanctuary side three rows with thirty-three pillars are discernible. Ultimately, it seems that this Arab conquest of Sind remained a local event without any great political impact—'an episode in the history of India and Islam, a triumph without result.' There seems, however, to have been a sustained cultural exchange, if it may be so called. The Arabs gained more from this initial contact than the conservative, almost perversely secretive and tradition-bound Hindus. While Arabs like Abu Maishar took advantage of the comparatively amiable relations between the Hindus and the Muslim Arabs to study Indian astronomy for over ten years in the distant Hindu holy city of Varanasi, the vainglorious Hindus, by and large, remained proud and aloof.

The Raids of Ghazni and Ghori

Whether the earliest footprints of Islam in India were destroyed by natural calamities such as the terrible earthquake of AD 893 or by the looting and conquest of the whole sea coast by subsequent Islamic raids of the Ghaznavids and the Ghorids is a moot point. However, the multiple invasions of the Turks and the Afghans proved to be of greater historical and political consequence than those of the early Arabs. Though the famous plunderer Sultan Mahmud of Ghazni did not prove to be an architect of empire, his numerous victories against the Rajput guardians of the north-western frontiers of India, particularly his audacious plunder of the legendary shrine of Somnath on the Kathiawar coast in AD 1026, laid bare the heroic but feeble defences of Hindu India. It is therefore, not surprising that the Ghor who succeeded the Ghaznavids continued to be lured by the riches of India. Mohammed of Ghori, descendant of the famous Ghorid Jahansuz (the 'world burner') having been appointed Governor of Ghazni, began his Indian campaigns in AD 1173. The earliest targets of this attacks were his co-religionists, 'the Islamic heretics of Multan and the Ghaznavid remains in the city of Lahore.' The undaunted Mohammed, in spite of tasting defeat once at the hands of Prithviraj Chauhan, the legendary Rajput ruler of Ajmer and Delhi, challenged him once again in AD 1192 on the historic battlefield of Tarain, the scene of his earlier humiliation.

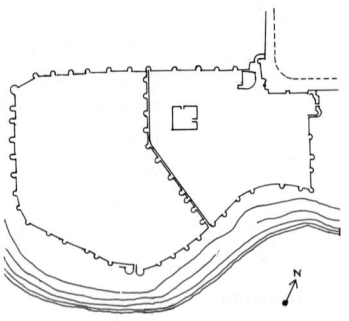

SITE PLAN

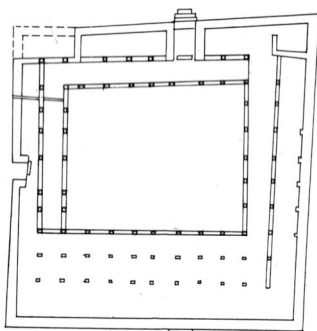

PLAN

Fig 1.01 Mosque at Bhambore, AD 727

'The Mohammadans spent the night before the battle telling their beads, and the Hindu army of Prithviraj Chauhan in listening to the heroic stories recited by their bards.' The battle that ensued was probably as fierce and in many ways as consequential as the legendary war of the *Mahabharata*. On that historic blood-drenched field Rajput valour and ponderous elephants could not prevail upon the mounted mobile archers of the Muslim armies. 'Like a great building the Hindu host tottered and collapsed in its own ruins'; ruins that were literally to become the foundations of a new and refreshing tradition of Islamic building ventures in India. Within an incredibly short period of twelve years since Prithviraj Chauhan's slaughter after the battle of Tarain, the Hindus of North India who, in Al Baruni's view, were 'too proud and self-centred to recognize the existence of outside nations', came under the rule of a Muslim slave king.

Qutb-ud-din and the Slave Dynasty

The spearhead of subsequent Muslim invasions of northern and eastern India was Qutb-ud-din Aibak, a Turkish slave of Muhammed Ghori. Muhammed had appointed him Viceroy of all his conquered territories. Qutb-ud-din took advantage of the moribund caste system of India, which designated the jobs of warfare and the government to the Kshatriyas, leaving other castes apathetic to the fate of the country. In rapid succession, he stormed Ajmer, the fortress of Gwalior, Delhi, Kanauj and ultimately even the distant Chandel stronghold of Kalanjar in Bundelkhand in AD 1203, Qutb-ud-din soon declared himself independent of Ghorid supremacy and wisely chose the Qila Rai Pithora built by Prithviraj of Delhi as his imperial capital. To Qutb-ud-din must go the credit of realizing that 'he who holds Delhi holds India.' Delhi, situated as it was at the mouth of the corridor between the foothills of the Himalaya and the fringes of the great Thar desert of Rajasthan, was the ideal location for defending India against foreign invasions along the north-west land route. Moreover, Delhi was the focus of the commercial and trading activity generated by the great hinterland of the Doab (literally 'two rivers') lying between the Ganga and the Yamuna. The first concern of Qutb-ud-din, though, was not with trade and commerce; he intended establishing the power of Islam over its newly subjugated people and proclaiming in concrete terms that with him Islam had come to stay and rule. The undisputed power of the sword of Islam had now to be consolidated through the efforts of its builders.

The Spirit of the Mosque

True to Prophet Mohammed's dictates of immediately installing a place of worship for the faithful on conquered territory, Qutb-ud-din decided to build a mosque to the everlasting glory of Islam in AD 1195. This was the Quwwat-ul-Islam (literally, the 'power of Islam') mosque within the fortified city of the Qila Rai Pithora *(Figs 1.02, 1.03)*. The decision to fulfil the Prophet's dictates was easier planned than implemented. Qutb-ud-din's Ghorid forces that invaded India consisted of soldiers, warriors and generals. Master builders, artisans and masons skilled in the art of building were naturally not a part of the army. The erection of a mosque, however, was imperative.

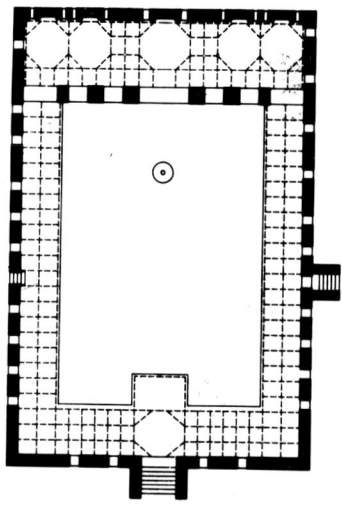

Fig 1.02 Plan of the Quwwat-ul-Islam mosque, Delhi

Fig 1.03 View of the Quwwat-ul-Islam mosque, Delhi

Moreover, it had to be constructed rapidly. Under the circumstances, the Muslims could not build in the country of their adoption without utilizing the skills of the indigenous artisans of the country. Thus, right at the inception of Islamic building activity in India, a sort of joint venture between the Hindu master builders and Islamic overseers was inevitable. Before the joint venture could get underway, the more knowledgeable of the soldiers and generals had somehow to explain the basic concept of a Muslim place of worship to the local builders. Fortunately, this was not a difficult task. Unlike the Hindu temple, mosque planning was not governed by complex geomantic theories of architecture. Rather, the basic concept of the mosque had evolved from the Prophet Mohammed's home in his birthplace. Here, a courtyard had been attached to his house to allow the faithful to gather. It consisted of a rectangular open-to-sky space, cordoned off by walls or cloisters. Within this court, the Muslim brethren could congregate and prostrate themselves to Allah. One of the logical essentials, therefore, of this otherwise austere and elementary requirement was that the worshippers be directed compulsively to pray in the direction of Mecca, as directed by Prophet Mohammed. This was achieved by the simple expedient of orienting the courtyard generally towards Mecca. The wall towards Mecca also had built into it the holy arch or *mehrab* and the *mimbar*, a sort of pulpit from which the priest would deliver the sermon. Thus, in India, it was the western wall that was critical to Islamic building requirements.

Vocabulary of Islamic Architecture

Apart from these rudimentary functional and religious requirements, Islamic architecture outside India had built up an identifiable structural vocabulary by the twelfth century. Though much of it was borrowed from the Roman and Byzantine systems, it had become a part and parcel of the Muslim architectural idiom. The two dominant elements of this language were the use of the pointed arch for spanning openings in walls, and the hemispherical dome for roofs. Although the faithful could well carry out the ritual of worship in a courtyard facing Mecca, it was felt that if this space could further be embellished with the familiar domes and arches it would add a feeling both of homeliness and pride to their custom of congregational prayers.

Some such intentions must have been conveyed by Qutb-ud-din's Muslim 'clerk-of-works' to the Hindu master builders. Given time, the stone-cutters would have been set the task of mining blocks of stone from the quarries. But time was not at hand. This was a period of constant warfare. Nevertheless, the faithful had to have their *masjid* (literally, a place of prostration), and the king, his mosque, as a monumental symbol, to impress upon the subjugated 'infidels' the power and rituals of the Islamic religion. So, if getting stone from quarries was a slow and cumbersome task, what was the alternative? To the straightforward militarily trained minds of Qutb-ud-din's Ghazis the solution was staring them in the face in the form of twenty-seven or so Hindu and Jaina temples that are said to have embellished Prithviraj Chauhan's Hindu capital. The destruction of these temples was more than a double-edged sword to Islamic intentions. Not only would it be a source of readymade building materials, but the demolition of the temples, apart from being a crushing blow to the already flagging spirit of Hinduism, would satisfy the iconoclastic zeal of the Sunni Muslim.

Temple Materials for Muslim Mosques

The structural sophistication of the Hindu masons who had built temples, facilitated the efforts of the demolition squads. They discovered the stone temples to be assembled out of meticulously cut structural elements such as beams, columns and lintels. These had been put together without mortar like a child's buildings blocks, but with the sophistication that modern technology employs to fabricate structures

Fig 1.04 Pillars from Hindu temples forming colonnade of the Quwwat-ul-Islam mosque, Delhi

out of precast concrete elements. Naturally, a building put together with so much precision could be ripped apart with equal ease and each element extracted intact. The only trouble with the building blocks recovered from the temples was that they were profusely covered with sculptures of the pantheon of Hindu gods and goddesses. Figurative representation was anathema to the orthodox and iconoclastic followers of the Sunni sect of Islam. In view of the expediency of the times, the Muslims had to be satisfied with defacing the sculpture by cutting off a nose here, an ear there or a face altogether in a rather feeble and crude attempt to render the human form unrecognizable. Hereafter, the task of erecting a mosque was fairly simple: re-assemble these structural elements to create a colonnade *(Fig 1.04)* around an open-to-sky courtyard, instead of the dark labyrinths of a Hindu temple.

Having chosen this blunt and expeditious method to erect their earliest mosque in India, the Muslim builders apparently baulked even at the idea of going through the cumbersome procedure of digging trenches and laying new foundations. What was easier than building over the existing foundations of the colonnaded corridors that surrounded the Hindu or Jaina sanctuary? And this, apparently, is what the Muslim overlords commanded them to do: raze the superstructure of the Hindu temple *in toto*, and re-erect the columns, beams and brackets over the existing foundations. Thus, the general east-west orientation of temples, too, suited their intentions admirably. To achieve their purpose of emphasizing the direction of prayers, all they had to do was to make the western veranda into a more spacious pillared hall. The rear wall of the hall so formed was then adorned with the traditional *qibla* arch to guide the faithful to pray in the direction of Mecca.

The Quwwat-ul-Islam at Delhi

The rudiments of a mosque measuring 217 ft× 150 ft (66 m × 46 m) built around a paved courtyard were quickly and easily assembled together by the builders of Qutb-ud-din. Whatever extra height was required, as at the corners, was achieved by superimposing one column over the other. No wonder then that Islam's earliest building effort in India, the Quwwat-ul-Islam mosque has been alluded to as an archaeological miscellany rather than as architecture. The description, however, is rather uncharitable. Even under these difficult circumstances, the Hindu builders put together an intelligible piece of architecture adorned with domes located over the corners of the courtyards and over the entrance canopy. These copolas were fashioned by infilling with lime concrete the terraced pyramid that would result out of the Hindu trabeate system of roofing, and then applying a thick layer of plaster to produce the smooth profile of a shallow dome *(Fig 1.05)*. Thus, the Muslim requisite of a dome was achieved without going through the structural obligation of building up a true dome.

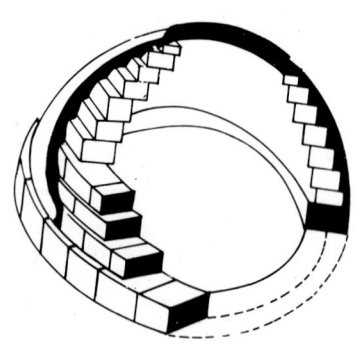

Fig 1.05

Fig 1.06

Fig 1.05 Hindu technique of constructing shallow dome

Fig 1.06 Hindu technique of constructing Muslim ogee arch, Quwwat-ul-Islam mosque

Screen of Arches for the Mosque

The stratagem of using the Hindu trabeate and bracketing system to make the shapes desired by Muslim conventions of building was repeated once again. Some years later it was felt that at least the western walls of the courtyard of the Quwwat-ul-Islam needed greater Islamic emphasis. To achieve this, the Delhi builders took their cue from the Caliph Osman who, as early as in the middle of the seventh century had erected a *maqsura* or screen of brick in front of the sanctuary in the Prophet's mosque at Medina. For erecting this wall in the Delhi mosque with five pointed arched openings, the builders had no Hindu precedence to go by. Prudently, they decided to

*Fig 1.07 Screen of arches,
Quwwat-ul-Islam mosque*

take no chances. This 108 ft (33 m) long stone masonry screen that was projected to rise to a height of 50 ft (15.2 m) was designed to be almost 8 ft (2.43 m) thick *(Fig 1.07)*. It is in the construction of the arches, the larger being a span of 22 ft (6.7 m) that the Hindu builder again resorted to his method of corbelling. He first created the rough multiple bracketed opening and then proceeded to chisel away the objectionable corners to create the smooth profile required of a Muslim ogee arch *(Fig 1.06)*. He was able to erect the arches mainly because they were more decorative than structural since they supported no load beyond their own. Otherwise, as had been common in Islamic buildings elsewhere, the true arch with radiating *voussoirs* would have had to be employed. But as we know, the Hindu builder had traditionally eschewed the arch on the rather feeble excuse that it never sleeps.

The next task in the completion of this project, though, was more in the ambit of the Hindu craftsman's true skills. The rough rubble masonry was covered with a veneer of red sandstone and the entire surface was decorated with rich patterns of carvings *(Fig 1.08)*. 'Some of the designs are the loveliest of their kind, particularly graceful being a border of spiral form having a floral device within each coil of its convolutions, emphatically Islamic.' Engraved upon the sandstone these have been described by a contemporary Muslim traveller 'as could not be done in wax.' But, of course, the Hindu stone-cutter had for centuries with infinite patience been treating even hard granite and marble as wax. Sandstone, therefore, was like butter to his chisel and these engravings in stone were but a mere flick of his finger. Unlike Persian Islamic architecture multicolour glazed tiles had been used to decorate the brick masonry. Muslim architecture in India continued to be influenced by the Hindu artists' penchant for more subtle monochromatic carved surfaces.

Fig 1.08 Rich patterns carved in red sandstone on the screen of arches, Quwwat-ul-Islam mosque

The Qutb Minar

The Hindu craftsman's talent for sculpture was soon put to a job more worthy of his mettle than the simple task of decorating plain surfaces of sandstone. In the year AD 1199 Qutb-ud-din Aibak, probably to celebrate his appointment as Governor, laid the foundations of a tower that was destined to rise to the staggering height of 238 ft (73 m). The confidence that Qutb-ud-din had in his Indian builders and designers is proven by the fact that 'though *minars* and towers were common features of Islamic architecture, particularly in West Asia, yet none of such size had been constructed.' Only a few years after making Delhi his capital, Qutb-ud-din embarked on his project. Even today, after eight hundred years of building in Delhi, and hundreds of architectural experiments, the Qutb Minar remains the most eye-catching monument in the capital. To many, Delhi *is* 'the city of Qutb' and archaeologists and historians are still breaking their heads to discover the sources that inspired the building of so radical a monument.

The Ghaznavid brick minarets near Kabul and Jam may well have served as models, but the Qutb turned out to be an edifice 'without parallel in Islamic art anywhere in the world.' Starting from its base of 47 ft (14.3 m) diameter, it tapers to a width of 9 ft (2.7 m), the top storey being approached by a central spiral staircase with 360 steps as it stands today (*Fig 1.09*). The verticality of its angular and curved outer turrets is broken by four balconies projected out over an elaborate system of stalactite-like pendentives (*Fig 1.10*). The sculptural treatment of the multiple niches below the balcony mark the Hindu carver's graduation in the art of non-figurative Islamic carving. It is true that the monument started by Qutb-ud-din and completed by his successor was also subsequently damaged by lightning, and had two extra storeys added in the Mughal period with a casing of marble. Ultimately, even the merloned balusters of its balconies were replaced according to the design of one Major Smith as late in AD 1828. Still, even as the Qutb Minar appears today with all its subsequent additions and alterations, it more than fulfils Qutb-ud-din's vision.

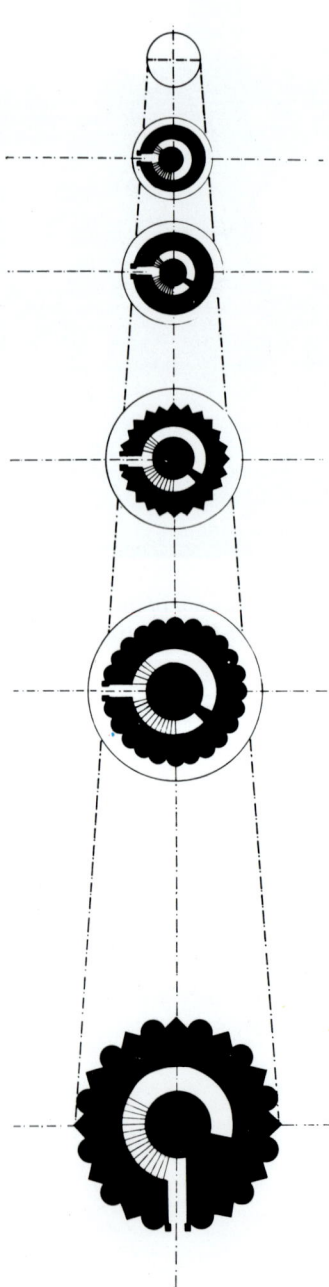

Fig 1.09 Plan of Qutb Minar

Fig 1.10 Pendentives supporting the balconies of the minar, Qutb complex

The Shadows of the Qutb

The intentions of Qutb-ud-din in building the Qutb Minar were certainly manifold. The Minar was first and foremost a tower erected to commemorate Muslim victories on the battlefield. It formed also an adjunct of the Quwwat-ul-Islam mosque in the Arabic tradition of towers attached to mosques like those of the Ibn Tulun at Cairo and the great mosque at Samarra. In these mosques, like the one at Delhi, the minaret from which the *azaan* given to call the faithful to prayer was an independent monument situated generally at the south-east corner of the courtyard. Once erected, the Qutb Minar also served the obvious military function of a watchtower. From its balconies, the plains below the scarp of the Aravalli range over which it was built were visible for miles around. To Islamic theology, though, 'the Qutb by its name signifies a pole and axis and thus the pivot of Justice, Sovereignty and of the Faith' that cast the 'shadow of God over the east and over the west.' It would seem that Yusuf I by raising the Giralda at Seville in Spain and Iltutmish by completing the Qutb at Delhi at about the same time were ensuring that the 'shadow of God' from the east and west should cover the entire Islamic world of the thirteenth century.

Mosque at Ajmer

The Islamic empire in India was still rather small. The furthermost centre of Muslim building activity under Qutb-ud-din's reign was at Ajmer in Rajasthan, barely 480 km west of Delhi. Here, too, the mosque that came to be known as the Arhai-din-ka-Jhonpra (literally 'two-and-a-half day cottage') was erected much like the earlier one at Delhi—an improvisation of columns, beams, and brackets recovered from Hindu and Jain temples. Only the scale of this mosque, built on a large platform cut out of the hillside is more impressive. In order to gain greater height for the sanctuary, as many as three of the standard columns of Hindu temples are superimposed. The roof of the prayer hall is a series of shallow corbelled domes planted over square

Fig 1.11 Screen of archways, Arhai-din-ka Jhonpra, Ajmer, AD 1205

pillared bays. Here, too, a screen of archways was at a subsequent period strung along the front of the *liwan (Fig 1.11)*. This screen, though more ornately built including trefoil, arches for the opening and minarets over the central arch, is not half as stately and majestic as the one at Delhi. It was built under the rule of Sultan Iltutmish, another able Turkish slave, perhaps a son-in-law of Qutb-ud-din. He had succeeded from the governorship of Badaun to the throne of Delhi through his ability and enterprise after Qutb-ud-din had died in an accident while playing polo. He faithfully completed many of Qutb-ud-din's ventures, both political and architectural, including the Qutb Minar as described earlier. After being victorious against the Rajput chieftains of Gwalior and Ranthambor and his own rebel governor of Bengal, Iltutmish found time to indulge in architectural schemes. In these he proved as successful as in his military and political exploits.

Expansion of the Qutb Complex

After completion of the grand Minar, it was obvious that the modest Quwwat-ul-Islam mosque was entirely overshadowed by the scale of this minaret. At the same time, with the growing Muslim population in the capital, a larger mosque was called for. However, Muslim power in Delhi had not yet become so all pervasive as to encourage Ilutmish to start a fresh building venture on a new site. He then made the astute decision to enlarge the existing mosque by throwing another colonnade, symmetrically arranged, around the existing one to create a courtyard that now included the Qutb Minar within its cloisters. The screen of arches was also appropriately extended along the northern and southern ends of the existing one *(Fig 1.12)*. Iltutmish had thus enlarged his place of worship to almost three times its original size by using a system akin to the method of throwing *parikrama* after *parikrama* around ancient shrines, which resulted in the great temple cities of Madurai and Rameswaram.

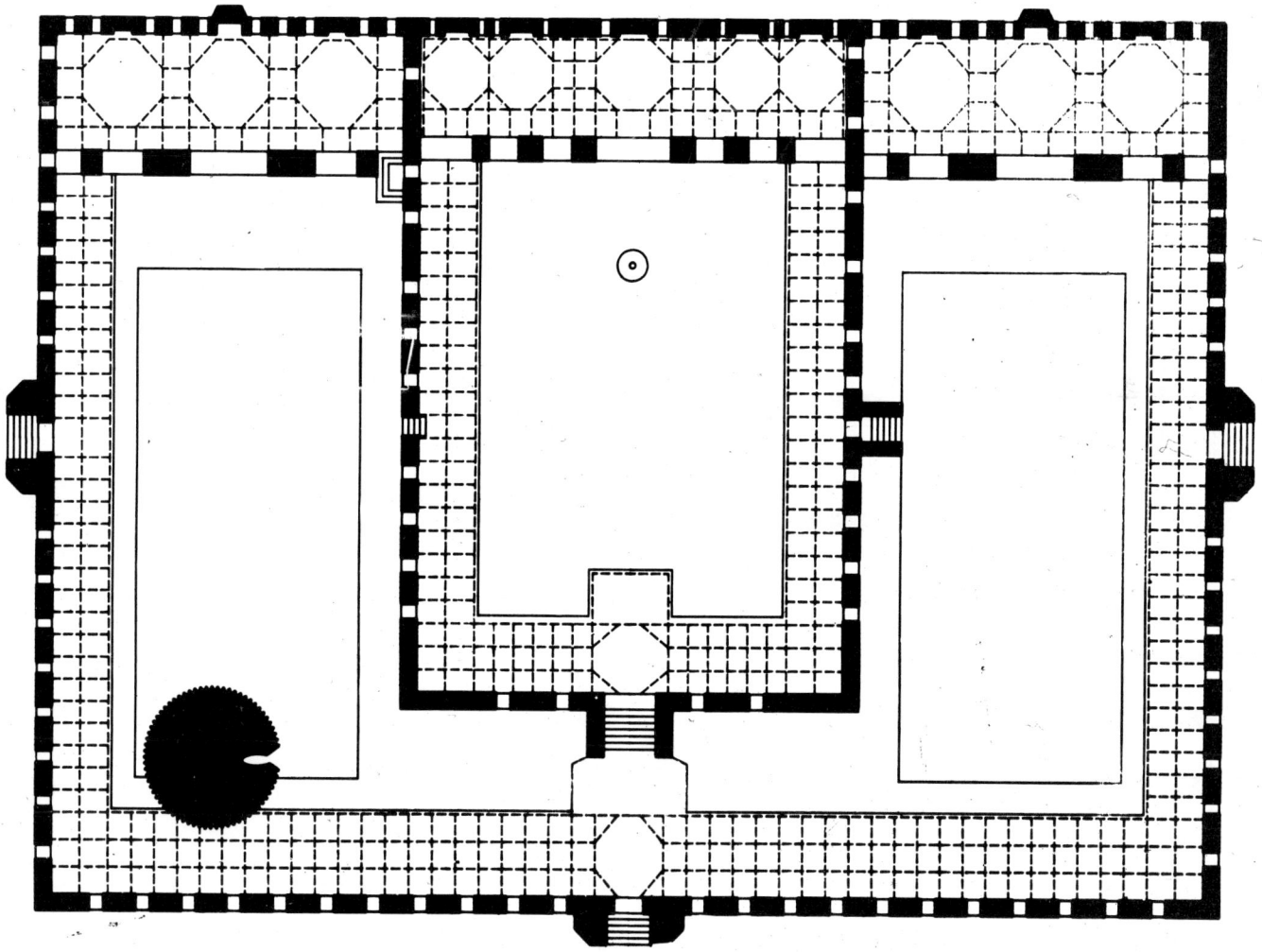

Fig 1.12 Enlarged plan of Quwwat-ul-Islam mosque by Iltutmish, AD 1229

The Cave of the Sultan

The significant part of Iltutmish's building activities was that of erecting the earliest Muslim tombs on Indian soil. The first, in AD 1231 is popularly known as Sultan Ghari or 'Cave of the Sultan.' It is a rather quaint non-traditional edifice for the tomb of his son Nasir-ud-din Shah, which is set in the middle of the courtyard of a small mosque *(Fig 1.13)*. The whole construction rests on a 10 ft (3 m) high plinth with domed circular towers making the corners of the courtyard. The grey masonry of the outer wall is almost cyclopean in scale and punctured by narrow arched apertures. The rather forbidding structure conveys the impression of a mini fort more than the 'tomb mosque' that it is *(Fig 1.14)*. The inside is altogether different. Here, the portico of the sanctuary along the western side is held up on, of all unlikely things, fluted white marble columns reminiscent of the Greek Doric order, altogether unprecedented in Indian architecture. The rest of the sanctuary is a reassembly of sandstone columns recovered from Hindu temples. The central portico of the *liwan* is crowned with a shallow corbelled dome that is octagonal in plan.

The centre of the courtyard of the mosque is taken up by the roof of the crypt of Nasir-ud-din that projects about 3 ft (0.9 m) above the level of the courtyard as an octagonal platform *(Fig 1.15)*. It seems that this platform at one time carried some kind of *chhatri* or canopy which has now vanished. The location of this rather unusual fortress-like structure about three miles (5 km) south-east of the Qutb complex (at present just off the road connecting Mehrauli to Palam) amidst a religious settlement of sorts, suggests that it served the manifold purpose of a tomb, a religious family shrine and, in an emergency, an advance army output for the capital.

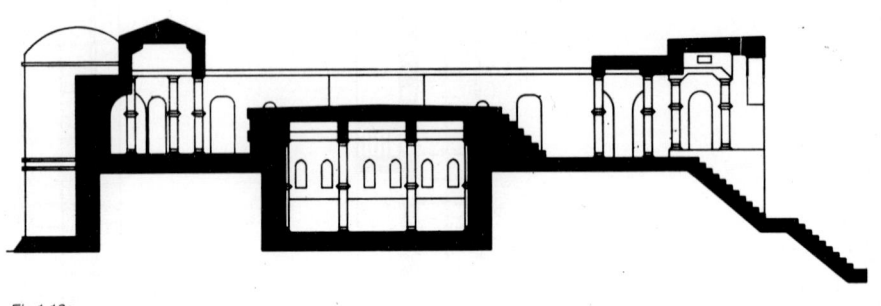

Fig 1.13a

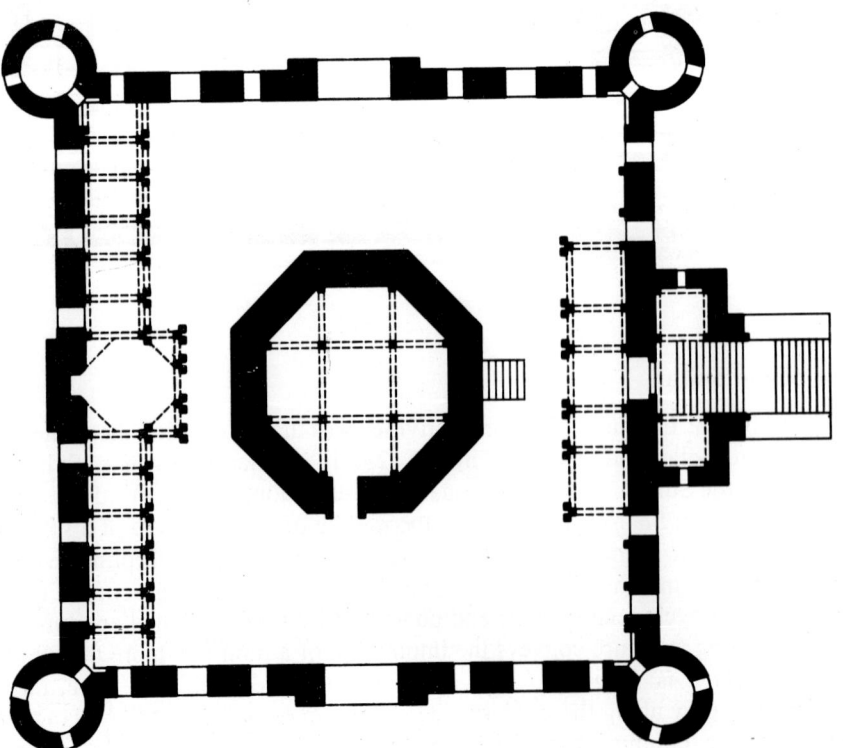

Fig 1.13b

Fig 1.13 Sultan Ghari's tomb, Delhi, AD 1231, (a) Section, (b) Plan

Fig 1.14 View of corner turrets of Sultan Ghari's tomb giving it a fort-like appearance

Fig 1.15 Portico of the sanctuary along the western side, Sultan Ghari's tomb

The Cube and Hemisphere of the Muslim Tomb

Sultan Iltutmish's own tomb was an altogether more comprehensible structure. It was built at the north-western corner of the Qutb mosque complex some four years after the construction of Sultan Ghari's tomb. Here, for the first time, the Indian builder came to terms with the most elementary grammar of the Islamic language of building. The crux of the Islamic structural language was the method of installing a dome that is essentially circular in plan over a cubic compartment that is imperatively square in configuration. Geometrically stated, this problem amounts to making the square (a polygon of four equal sides) and the circle (a polygon of infinite sides) approximate each other; structurally interpreted, it amounts to supporting the circular drum-like base for the springing of the dome over the walls of the square compartment below

(Figs 1.16). The geometrical solution lay in increasing the four sides of the square to eight of an octagon, and progressively to sixteen- and thirty-two-sided polygons gradually approximating the infinite sides of a circle *(Figs 1.17, 1.18).* The corresponding structural solution lay in effectively spanning the right-angular corners of the square to create an octagon, then spanning the obtuse corners of the octagon, and repeating the process until a circular ring was achieved. The first constructional step then, was to put an arch across the corner, unless the span was small enough to be bridged by a simple stone beam. Then the process of spanning the corners could be repeated until a circular ring of masonry could be safely projected out over the lower polygon. Over this circular base, the dome of the desired profile could be erected. A cubic base pierced by arched openings crowned with a hemispherical dome *(Fig 1.16)* is the most essential and elementary structural and visual unit of almost the entire gamut of Islamic architecture, be it tombs, mosques or palaces in India, Arabia, Turkey, Persia or anywhere.

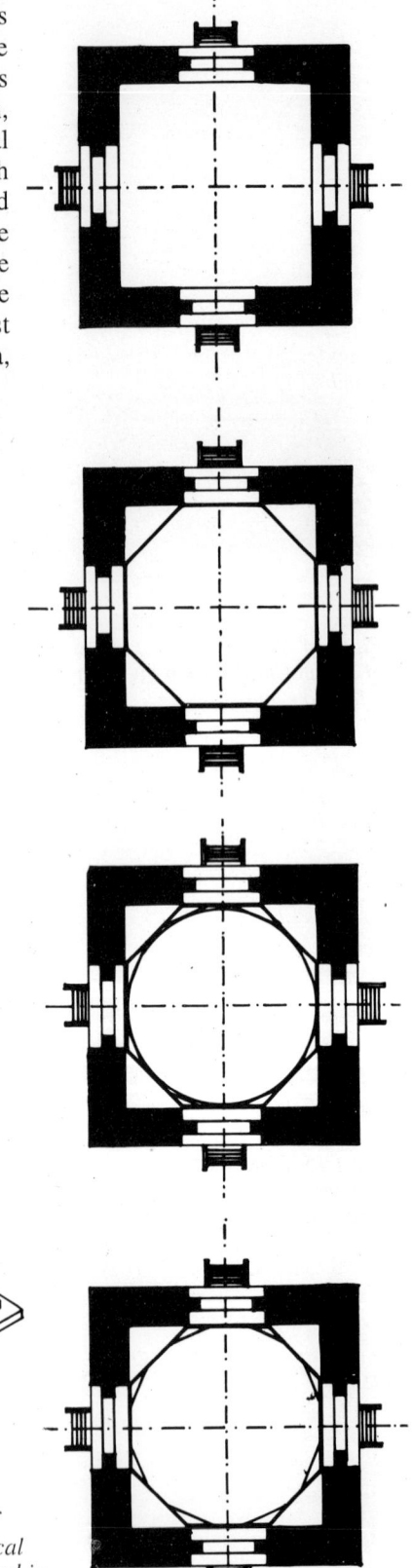

Fig 1.16 A cubic base pierced by arched openings crowned with a hemispherical dome

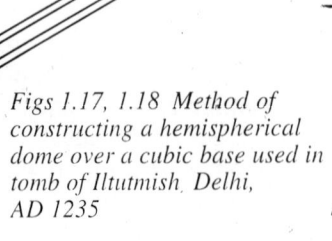

Fig 1.17

Figs 1.17, 1.18 Method of constructing a hemispherical dome over a cubic base used in tomb of Iltutmish, Delhi, AD 1235   Fig 1.18

The Tomb of Iltutmish

In employing this new grammar for the first time, the Indian builder, true to his innate genius, was more successful in the decorative aspects of Iltutmish's tomb than in the structural aspects. He managed to build the initial arch across the right-angular junction corners of the walls below *(Fig 1.19)*. These arches, in turn, were covered with highly intricate and beautiful Islamic arabesques delicately sunk both into sandstone and marble panels. With their method of corbelling by building oversailing courses of masonry, they were, however, unable to put together a dome that could stand the test of time *(Fig 1.20)*. Today, we are left only with the square base of this tomb, the 30 ft (9 m) diameter dome having collapsed long ago. It was apparent that the Hindu master mason, whatever his inherent restraints, had sooner or later to come to terms with the building of the true arch and dome if he were to create architecture for his new Muslim overlords.

Fig 1.19 Detail across the right angular junction of wall corners, tomb of Iltutmish

Fig 1.20 Section of the tomb of Iltutmish

The True Arch and the Tomb of Balban

Ultimately, this was done, however timidly, in the tomb for Balban, formerly a Turkish general in the army of the Slave Dynasty kings. In the later years of his life, at the end of the thirteenth century, Balban provided some stability after the turbulent state of affairs that had prevailed under the rule of Sultana Raziya. Sultan Iltutmish, in an unprecedented manner had nominated his daughter Raziya, as his successor, who by some accounts was 'endowed with all the admirable attitude and qualification necessary for kings.' She was nevertheless unable to control the machinations of the feudal governors and generals who surrounded her. She made herself even more vulnerable not only by appointing a black Abyssynian slave, one Jalal-ud-din Yaqut as a general, but also, by some accounts, accepting him as her lover. Being preoccupied with her turbulent political and romantic life, she had little time left for building activity. Thus, the Indian builders had to wait for a period of over 40 years until the death of Balban in AD 1287 to undertake their critical experiments in Islamic building.

Fig 1.21 True arches, tomb of Balban, Delhi, AD 1287

Fig 1.22 Ruins of tomb of Balban

  Balban's tomb, though modest in size, is nevertheless of radical to Islamic building history in India for, Percy Brown puts it, 'here for the first time a true arch was put together (in India) and bonded on the scientific system originally formulated by the Roman engineers' *(Fig 1.22).* Of this tomb, too, like that of Iltutmish's little more than the structure of walls has survived. To the evolution of Islamic architecture in India it is a building of greater significance than even the Qutb. For the unobtrusive true arches built with radiating voussoirs *(Fig 1.21)* symbolize the confidence with which subsequent Islamic dynasties were to continue to embellish the subcontinent with an architecture that was essentially Indian, and in its aesthetic and structural elegance could challenge the best of Islamic architecture anywhere in the world.

Kali Masjid,
Shahjahanabad

CHAPTER 2

The Khaljis and Tughlaqs of Delhi

AD 1290–AD 1413

With the end of Balban's dictatorial rule, 'the stability of which depended upon the personal strength of the ruler', Delhi subsequently was ruled by a number of wayward and ineffectual kings, including Balban's grandson Kaiqubad. He 'plunged himself at once into a whirlpool of pleasure' at his favourite Surkh Mahal or Red Palace on the banks of the Yamuna in the newly built settlement of Kilokri. Kaiqubad was soon reduced to a physical wreck and done to death by a Khalji noble whose father had been executed at Kaiqubad's behest. The Khalji people, though originally of Turkish origin, had acquired a deep overlay of Afghan characteristics due to their long tenure in the reign of Khalji near Ghazni in Afghanistan. There was no love lost between the so-called 'Turkish party' of the Slave kings and the Khaljis who were struggling to establish their supremacy at Delhi since the death of Balban.

Ascent of the Khaljis

Ultimately, in AD 1290, Firoz Jallal-ud-din Khalji, at the ripe old age of seventy ascended the throne of Delhi. Jallal-ud-din's was a brief rule of six years, during which 'preoccupied as he was with preparations for the next world,' he showed 'the most impolitic tenderness towards rebels and other criminals.' He devoted greater energies to the pious duties of a good Muslim than to statecraft worthy of a monarch, and was eventually unable to hold the reins of power in those troubled times. His only success seems to have been against a horde of Mongols who, under a grandson of Halaku, invaded India in AD 1292 and were convincingly defeated by the Khalji armies. Even 'towards the captured Mongols, Jallal-ud-din showed leniency and unwisely allowed many of them to set up their camps just outside Delhi and live there as the so-called 'new Mussalmans.' Ultimately, it was not the dreaded Mongols but Jallal-ud-din's own ambitious nephew Ala-ud-din who ended the peaceful life of his uncle by the familiar pattern of deceit and treachery. Ala-ud-din murdered Jallal-ud-din when the old man came unarmed out of the city to welcome him back from his triumphant ventures in Devagiri in the South. Soon after, he had himself crowned at Delhi by winning over the nobles through a lavish distribution of the 'Deccan gold' that he had plundered during his invasion of Devagiri.

Ala-ud-din, the Great Khalji

Once established on the throne, Ala-ud-din continued his series of successes as military commander. Apart from warding off numerous Mongol invasions of India, he 'stamped out the last embers of Hindu rule' by annexing Gujarat and Ranthambor and the great fortress of Chittor. According to legend, securing the beautiful Princess Padmini was one of the chief aims of Ala-ud-din's siege of Chittor. Ala-ud-din Khalji proved to be a tremendously spirited ruler. He seemed to do nothing in half measure, being as intensely pious as he was fiendishly cruel. 'He shed more innocent blood than even Pharoah was guilty of.' Yet, Ala-ud-din's court at Delhi attracted a large number of such Muslim luminaries as the poet Amir Khusrau, and the historian Baruni. His early military successes turned his head, and he soon developed the attributes of a true megalomaniac. He had himself inscribed as Alexander II on the coins minted in the Delhi treasury. His architectural projects, too, seem to reflect his character. It was during his powerful rule after a torpor of over three quarters of a century, that Islamic builders took a definitive step forward in their art, venturing into many a grandiose and foolhardy project. Thus, the Alai Darwaza at the Qutb, the large water tank of the Alai Hauz (later Hauz Khas), and even the frugal remains of his new city of Siri bear testimony to his dynamic ability. The 70 ft (21.3 m) high rubble stump of a minar that was envisaged to rise higher than the Qutb, is equally reminiscent of his vainglorious dreams of being another Alexander.

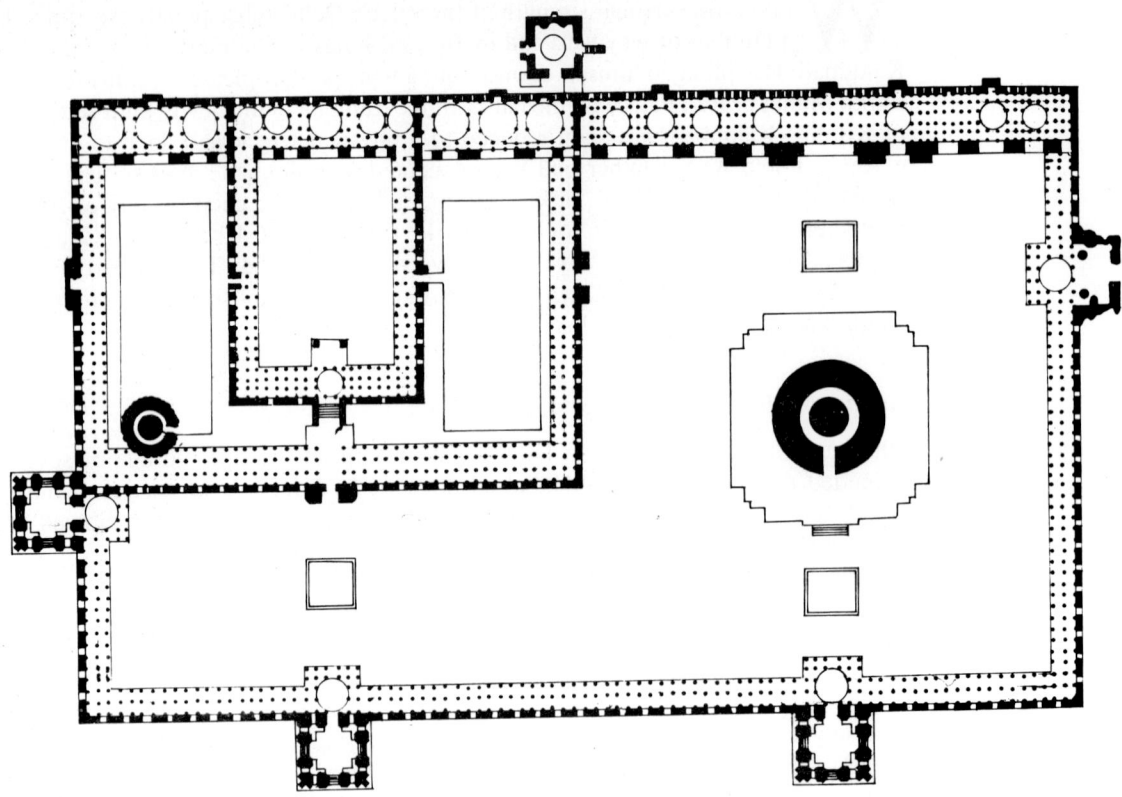

Further Extensions of the Qutb

The Alai Darwaza and the remains of the so-called Alai Minar, as they stand today, were also only a minor part of a much grander scheme. While Iltutmish had been content with increasing the size of the Quwwat-ul-Islam mosque by three times. Ala-ud-din further enlarged it by more than six times. This was achieved by throwing yet another asymmetrically arranged cloister around the existing one *(Fig 2.01)*. In the centre courtyard of the extension on the northern side, he laid the foundations of the Alai Minar. At symmetrical intervals along the outer walls, it was proposed to install six gateways *(Fig 2.02)*. Of these, either the Alai Darwaza was the only one to be completed, or is the only once to survive to date. The Alai Darwaza, though modest in size, marks the beginning of the process of refinement of the 'basic module' of Islamic architecture — the cube and the hemisphere — as assembled for the first time

Fig 2.01 Plan of the Qutb complex, Delhi, including extensions carried out by Ala-ud-din Khalji

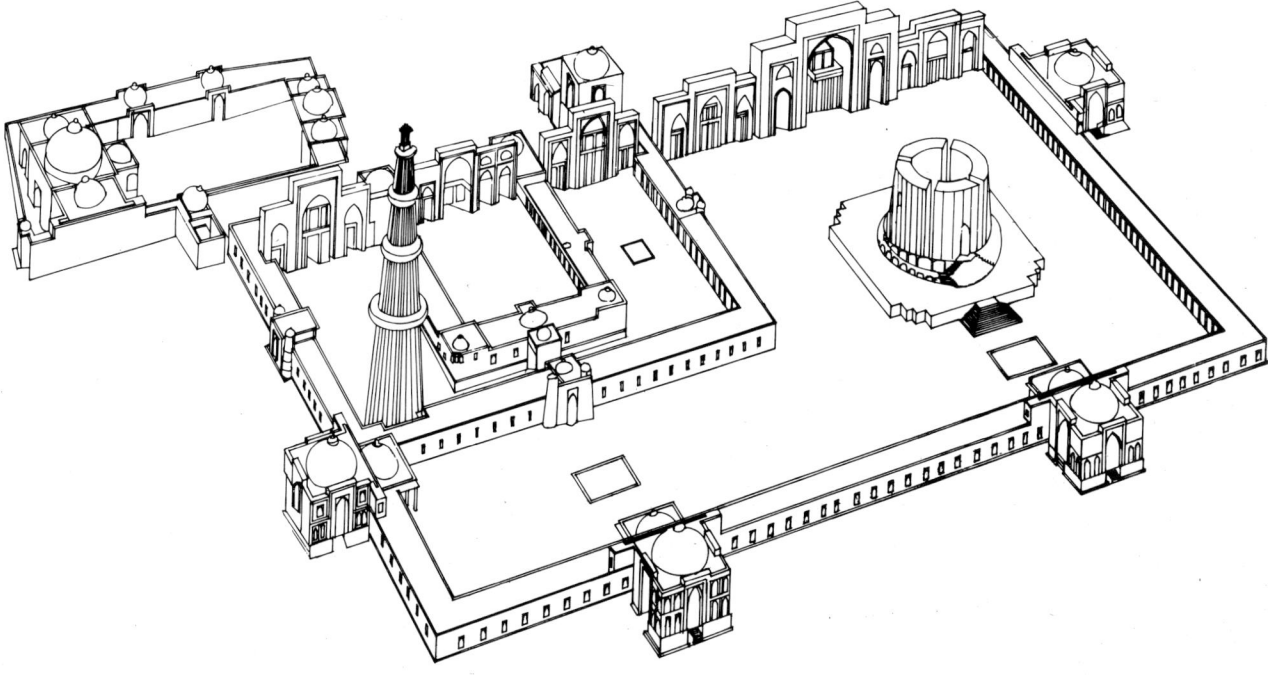

Fig 2.02 View of the Qutb complex

in Delhi in Iltutmish's tomb. The refinement is apparent both in its structural and decorative techniques. For the first time, the construction was carried out with masonry that was formed of alternate courses of stretchers and headers, with the headers embedded deep into the thickness of the wall, ensuring greater stability to the structure. What is more, the arches, the squinches *(Fig 2.03)* and the dome of 34 ft (10.3) diameter over the 56 ft (17 m) square base are all constructed with true arches as seen for the first time in Balban's tomb over seventy-five years ago.

The Red and White Alai Darwaza

To the lay viewer not too concerned with these technical niceties the treatment of the three outer sides of the Alai Darwaza's cubic substructure is a visual delight *(Fig 2.04)*. The most invigorating aspect of this is the refreshing choice of a blend of red sandstone and white marble as facing materials. Into these two materials, the Indian carver effortlessly stencilled the flat Quranic inscriptional bands to surround and define the openings, and took the liberty of applying more sculpturesque indigenous details where appropriate. Thus, while the horse-shoe shaped arch is defined by hands of inscriptions in marble, the jambs are adorned with pairs of slender pilasters undoubtedly inspired by Hindu temple columns. The intrados of the arch itself is ornamented with the so-called 'spear head fringe' or 'garland of buds'. This device may be interpreted either as a miniaturization of the Jaina *torana*, or as a subdued amplification of the geometrical ornamentation permitted by Islam. Either way it became a popular technique for softening the stern profile of the pointed arch. In broader terms, the treatment of the outer facade as a two-storeyed building with blind windows in its non-existent upper 'storey' proved an enticing technique and became the standard Islamic method of visually reducing the volume of the lower cubic mass to more comfortable human proportions *(Fig 2.04)*. With infinite patience, the builders have lavished the interior, too, with equally intricate arabesques. Seen in the light that filters in through them the star and hexagon *jaalis* that fill the arched windows in the lower 'storey' create intricate and complex patterns. These grills were the forerunners of what the Mughals were later to elevate to great works of art. The *jaali* was an eminently sensible architectural device to provide controlled illumination and ventilation for the large voluminous spaces that were desired by Islam in contrast to the small and dark cubicles of the Hindu temple.

Fig 2.03 View of the sandstone and white marble facade of Alai Darwaza, Qutb complex, Delhi

The Jamat Khana Masjid

The first of these large spaces to be produced by rationally conjoining together three cubic compartments each with its own dome, was assembled in the building of the Jamat Khana (literally, 'place of congregation') Masjid near the Dargah of Nizam-ud-din Auliya *(Fig 2.05)*. This *masjid*, though rather austere in comparison to the Alai Darwaza, nevertheless marks a major evolution in the design development of mosques in India. Here, for the first time, the rather prosaic *liwan* is transformed into one composite rectangular hall, uninterrupted by columns. What is more, the three arched openings in its eastern wall satisfactorily replace the *maqsura* or attached screen of arches. The span of the arches, too, is clear indication of the Indian builders' growing familiarity with the construction of the true arch.

Ala-ud-din's efforts to live up to his Alexandrian image through his building ventures were indeed ambitious. In addition to the ill-fated Alai Minar *(Fig 2.06)* these included the building of the circular city of Siri (near modern Panchsheel Enclave) 'that had seven gates', and the Hauz Khas, so large that 'an arrow cannot

Fig 2.04 Corner squinch of the Alai Darwaza, Qutb complex, Delhi, AD 1305

Fig 2.05 View of Jamat Khana Masjid, Delhi, AD 1320

Fig 2.06 View of Alai Minar, Delhi

be shot from one side to the other.' The only notable architectural venture of this period was the Alai Darwaza. History seems to have paid back Ala-ud-din in his own coin. The grand city of Siri, in time-honoured tradition, was ravaged by later emperors to provide material for their own dream cities, and the Hauz Khas as it stands today was virtually rebuilt by the later Tughlaq kings. Ala-ud-din died a pathetic death as a pupet in the hands of Malik Kafur, his Chief Minister, and even the identity of his tomb among the rather nondescript structures just behind the Qutb mosque is doubtful.

The Prolific Tughlaqs

The end of Ala-ud-din's stern rule marked the beginning of the usual series of conspiracies, treachery and cruelty, including the blinding and slaughter of royal offspring. The ironic results of this Turkish blood bath was that a supposed Hindu convert, Khusrau Shah, captured the throne and outdid his Muslim masters in carrying out a large scale massacre of the late Sultan's friends and relatives. He is said to have distributed largesse and favours to the otherwise impoverished Hindus. The Hindu resurgence, though, was brief and somewhat inconsequential. Within exactly four months and four days of his installation as ruler, Khusrau Shah's armies were unceremoniously defeated by one Ghazi Malik, Governor of Dipalpur, near the city of Multan in north-western India. To the great relief of and with the connivance of the harassed true blood Turkish and Afghan nobles of Delhi, Ghazi Malik ascended the throne of Delhi as Sultan Ghias-ud-din, and established one of the most prolific building dynasties of Delhi, that of the Tughlaqs. Ghias-ud-din's rule was but a brief one of just five years, but not entirely as inconsequential as that of Jalal-ud-din Khalji's. The tempo for the unceasing building activity of the Tughlaqs was initiated by Ghias-ud-din's decision to build a new fortress city.

Architectural Inspiration from Multan

Ghias-ud-din's 'former fiefdom of Dipalpur lay directly on the classical route of western invasions into India,' and he had become the traditional 'warden of the marches.' Hence, he had developed a distinctive notion of architecture, more akin to that of fortresses and military establishments than to seraglios of pleasure and places of worship. Added to this was Ghias-ud-din's familiarity with Muslim architecture in Multan that was composed largely of brick. A typical example of this is the tomb of Rukhn-i-Alam built within a fortress-like structure in Multan in AD 1340 *(Figs 2.07a, b, c)*. In the building of a massive structure like the Rukhn-i-Alam with brick, which has a lower crushing strength than stone, it is imperative that the load of the superstructure be gradually reduced as it rises higher to prevent the bricks in the foundations from being pulverized. Thus, the most dominant feature of the architecture of Multan was the characteristic tapering brick masonry walls that were gradually reduced in thickness as they rose higher. In erecting the outer ramparts of the new city of Tughlaqabad in stone masonry the builders seem to have taken their cue as much from Ghias-ud-din's architectural inclinations as from the contours of the hill over which they were building. The walls of the city became massive retaining walls of masonry that virtually hug the hill slopes on which they were built *(Fig 2.08)*.

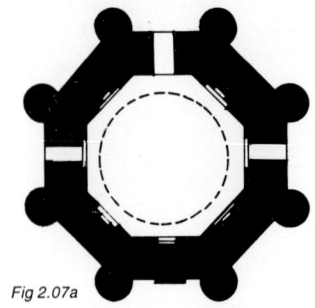

Fig 2.07a

Fig 2.07 Tomb of Rukhn-i-Alam, Multan, AD 1340, (a) Plan, (b) Cross-section (c) View

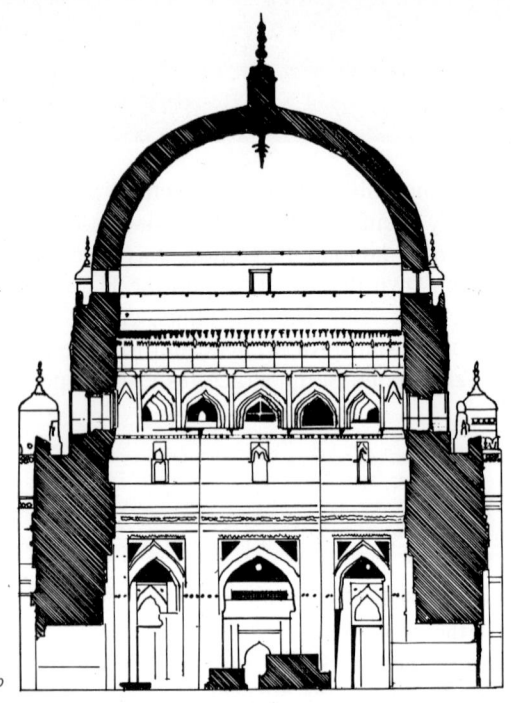

Fig 2.07b

Fig 2.07c

Fig 2.08

Fig 2.09 Sketch plan of the tomb of Ghias-ud-din Tughlaq, Delhi

Fig 2.10 The tomb of Ghias-ud-din Tughlaq, Delhi, AD 1325

Fig 2.08 Masonry fortress walls of the city of Tughlaqabad, Delhi (facing page, below)

Tomb of Ghias-ud-din

The builders of empire seem to have been profoundly enamoured by the powerful militant appearance of the fort of Tughlaqabad. So much so, that the building of conventional vertical walls for even major monuments was abandoned. Buildings were constructed with walls that made an angle of as much as 75° with the ground, instead of rising up in plumb. This, 'though little can be said of the ravaged but once grand bazaars, mosques and palaces of the inner city of Tughlaqabad,' much can be deduced of the Tughlaq style of architecture from the well-preserved tomb of Ghias-ud-din. This tomb of the founder of the Tughlaqs was built in AD 1325 within an unusual irregular pentagonal fortified enclosure. The plan, no doubt, was dictated by the contours of the hillock just outside the southern gates of the fortress over which this barbican-like structure was planted *(Figs 2.09, 210)*. The tomb is connected to the fortress by a 250 yd (228.6 m) long causeway, built over what at one time must have been a large sheet of water, but today is dry scrubland. In its intent and function, as a fortified tomb outside the city, the complex is not unlike that of the Sultan Ghari described earlier.

Fig 2.11

Fig 2.12

Fig 2.11 Use of true arch with lintel in openings of Ghias-ud-din Tughlaq's tomb, Delhi

Fig 2.12 Kalasa topped dome of tomb of Ghias-ud-din Tughlaq, Delhi

The Arch and Lintel

Apart from the distinctive 75° camber of its outer walls, the 61 ft (20 sq m) square tomb of Ghias-ud-din also heralds various other distinct features that were to continue to nourish Islamic architecture in India for centuries to come. The most characteristic of these is the mixed attitude of the Hindu builder to the arched form of construction and the lintel and beam method. In the tomb of Ghias-ud-din Tughlaq they struck a peculiar form of compromise. In spite of using the true arch to span the openings, a redundant stone lintel was installed just below the springing of the arch *(Fig 2.11)*. Whether the insertion of the lintel was the Indian builder's way of saying 'Yes, the arch is alright, but there's no harm in adding a lintel to ensure stability,' or whether it was a practical solution to being able to install a rectangular timber door in the arched openings is a moot point. But it is certain that this architectural 'compromise' became an elegant and effective device in the building style of the Tughlaqs as well as their successors.

Kalasa over the Muslim Dome

One of the symbolically rich Hindu elements that crept into Tughlaq architecture without a murmur of protest from the devout Muslims was the elegant *kalasa* pinnacle. It was planted at the apex of the Tughlaq dome to replace the rather ungainly 'nipple' that had marked the culmination of the shallow dome over the Alai Darwaza. It is a fact of history that unlike the renaissance architects who crowned their domes with an appropriate lantern of sorts, Islamic architecture never quite evolved an appropriate pinnacle for its dome. With its bare crest this seemed incomplete to the cultured eye of the Indian builder. He then adorned it with a duly modified version of the *kalasa* pinnacle with which he had completed hundreds and thousands of *shikharas* of Hindu temples. The tomb of Ghias-ud-din Tughlaq capped by a marble encased and fully contoured *kalasa* topped dome, rising to a height of 80 ft (24 m) over the merlon fringed pyramidical base *(Fig 2.12)* presents an altogether more resolute and distinctive form of Islamic architecture in India than the more elegant and pretty Alai Darwaza.

Mohammad Tughlaq, The Prince of Moneyers

The burial of Ghias-ud-din Tughlaq (in a tomb built for himself), when he was murdered by his son Juna, proved symbolic of the architectural future of India under his son. Ghias-ud-din was killed by his own son by a novel method of destructive engineering. A huge canopy that was erected to welcome him to the city, collapsed over his head just at the moment when he was alone under it. Juna immediately had himself crowned as Muhammad-bin-Tughlaq and indulged in many more projects which unintentionally turned out to be equally destructive. Mohammad was a learned and devout king keenly interested even in medical sciences and philosophy. Yet, he remained an uncanny mixture of opposites at whose door 'is seen always some pauper on the way to wealth, or some corpse that has been executed.' It is thought by many that some of his experiments such as floating a copper currency and practising secularism and impartial justice proved disastrous only because these were far ahead of his times.

Judging from the scanty remains of his building projects, his architectural ideas, however, show no signs whatever of being ahead of their time. The only extant complex that can be attributed to his period is the so-called Bijai Mandal (near modern Panchsheel Colony), part of a palace complex whose structures merely indicate that the Ghias-ud-din style of building was persisting. Muhammad Tughlaq linked together the fortress of Tughlaqabad and the Qutb complex to create his new city of Jahanpanah. He also added the fort of Adilabad at the other end of the fortified dyke of the lake opposite Tughlaqabad to make Delhi 'a great city rivalling Cairo and Baghdad in size and prosperity.' But then, without batting an eyelid, he ordered the entire population to shift the camps of his new capital at Daulatabad in the south

Fig 2.13 Daulatabad Fort, AD 1334

(Fig 2.13). His soul was content and mind at rest only when 'gazing upon Delhi, he found no fire nor smoke nor light' and 'not a cat or a dog was left among the buildings.' Ultimately, 'the king was freed from his people and they from their king,' when Muhammad Tughlaq died a sudden death while campaigning in the region of Sind in an effort to consolidate his disintegrating empire.

Firuz Shah Tughlaq, The Prince of Builders

Firuz Shah, who succeeded Muhammad Tughlaq, to whom was bequeathed a depleted empire, must have had to make heroic efforts indeed to fulfil his divine duties. He has claimed in his writings that 'among the many gifts which God bestowed upon me, His humble servant,' was a desire to erect public buildings. One may well take with a massive pinch of salt the claim of historians 'that he made 1,200 gardens around Delhi,' and 'is credited with the erection of two hundred towns, forty mosques, thirty villages, thirty reservoirs, fifty dams, one hundred hospitals, one hundred public baths, and one hundred and fifty bridges.' Nevertheless, judging from the extant remains, there is little doubt that Firuz Shah's building efforts were indeed extensive, even though spartan in appearance. Austerity, in terms of finish and constructive methods, remained the hallmark of this prolific period of building. No wonder, since Firuz Shah, the Tughlaq 'prince of builders' succeeded Muhammad, the so-called 'prince of moneyers' who had bankrupted the Delhi treasury by his experiment with the finances of the empire. Firuz Shah's thirty-seven years of experiments with architecture proved more successful than those of his predecessor with the economy. The confidence with which Firuz Shah set out to give concrete shape to 'the gift that God bestowed on him' is apparent from the choice of his first building venture in Delhi, that is, the erection of a new capital city, and that too, on virgin territory well away from the earlier sites of the Khaljis and Tughlaqs which were located near the Qutb region. His was the first Islamic city of Delhi to be built on the banks of the River Yamuna.

The Cities of Delhi Move North

It would appear that the source of water supply around the Qutb area was no longer reliable. Also in these less turbulent times, the need was no longer felt for building gargantuan fortess structures over the hills of the Aravalli range. In choosing to build on the banks of the Yamuna well north of the earlier three cities of Delhi, Firuz Shah set up a precedent that was to be followed for centuries. Every new city of the so-called 'seven cities of Delhi' was invariably built north of its predecessor. The reasons for this are not difficult to understand. The builder of a new city could naturally want to receive both fresh air, and the waters of the Yamuna, uncontaminated by the dying remains of an old city. And both the fresh winds and the Yamuna flowed into Delhi from the northern mountain ranges. In fact, a similar rationale must have prompted the Khaljis and Tughlaqs to build north of the then existing city.

Fig 2.14 View of the citadel of Firuz Shah Kotla, Delhi, AD 1354

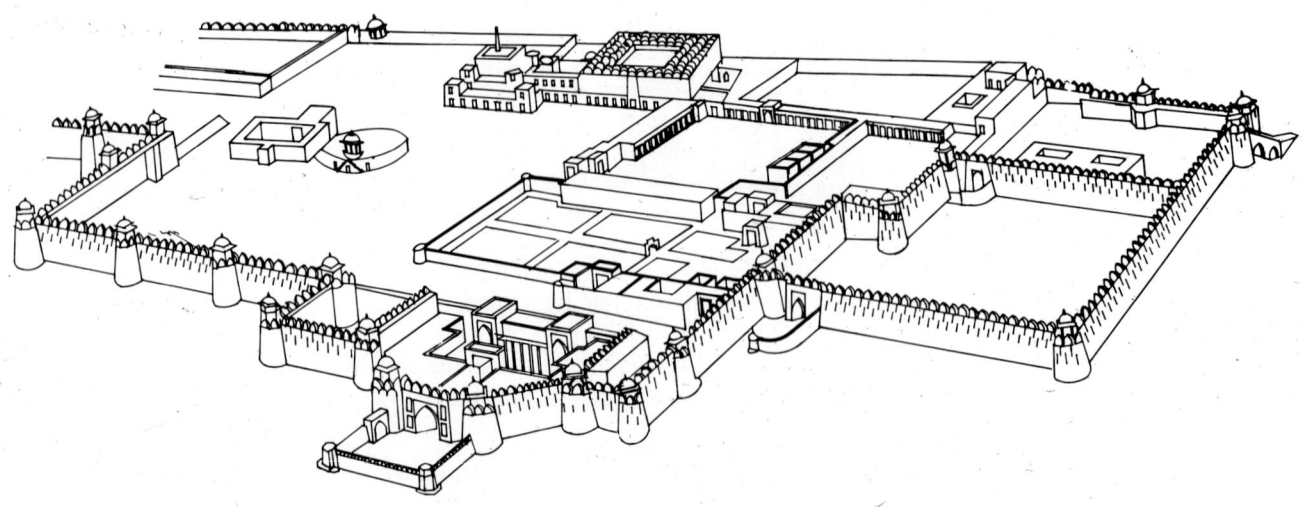

Fig 2.15 Ruins of the defensive walls of Firuz Shah Kotla, Delhi

The City of Firuz Shah Kotla

Apart from making a more rational choice of the site, Firuz Shah, in the planning of his new city, set up precedents that remained valid for centuries. His city palace, in fact, became the prototype of the great Mughal palace cities of the sixteenth century. These planning standards were possible, no doubt, because the Muslim administration of India was gradually jelling into a discernible pattern that could be interpreted in terms of town planning. The Firuz Shah Kotla, as the palace city of Firuz Shah Tughlaq came to be known, is planned as a slightly irregular rectangle, half a mile (800 m) long and a quarter mile (400 m) wide, defined by moderately defensive walls *(Figs 2.14, 2.15)*. The eastern and longer side of the rectangle was parallel and abutted the banks of the river. In the middle of the side opposite the river was the main entrance gate planned in the usual manner of a protective barbican. Directly opposite this was a large rectangular court defined by cloisters, meant to be the Diwan-i-Am (hall of public audience) where the King, when in residence, would daily give audience to the common public. Just behind this was the Diwan-i-Khas (hall of private audience) where the King held his 'cabinet meetings' and met the important officers of his administration. Right along the river banks that were not only comparatively safe against military attacks but also afforded the finest view were located all the private palaces, mosques and the harems of royal court. The areas north and south of the central axis were divided into various square and rectangular courtyards, in which were a 'great variety of structures such as pavilions for different purposes, grape and water gardens, baths, tanks, barracks, armoury and servants' quarters all conveniently disposed and communicating with one another.' The broad planning principles of a Muslim 'city centre' enunciated for the first time by Firuz Shah consisted in locating the Diwan-i-Khas at the heart of the complex, backed up by a series of private palaces along the river front, well protected on the other side by army barracks and other sundry structures, with limited access for the common public to the Diwan-i-Am.

'Militant' Palaces and Mosques

Firuz Shah's builders proved to be just as innovative in the detailing of some of the palaces within the complex as in the hundreds of mosques built during his reign. Within the Kotla is located a curious three-tiered structure, every platform of which is set back from the preceding one to create terraces in front of the series of compartments on the periphery *(Figs 2.16a, b)*. No doubt designers of this building were inspired by the terraced Buddhist *viharas* that abounded in the Indian countryside. The structure was most probably inhabited by the numerous concubines of the King. The picturesque quality of this indigenously inspired structure is appropriately completed by a stone *sthamba* of the Asoka period planted at its apex *(Fig 2.17)*. The latter is believed to have been brought here from its original site somewhere near Ambala. The most characteristic of the numerous mosques of this period in Delhi are the Kali Masjid (AD 1370) at Nizam-ud-din Auliya, the Kalan Masjid (AD 1375) in Shahjahanabad, and Khirki (AD 1375) and Begumpura (AD 1370), Masjids in the vicinity of modern Malviya Nagar. An architectural trait common to all these is the persistence with the 'pseudo-militaristic' style of Ghias-ud-din in the structuring even of religious edifices. This, however, was not achieved

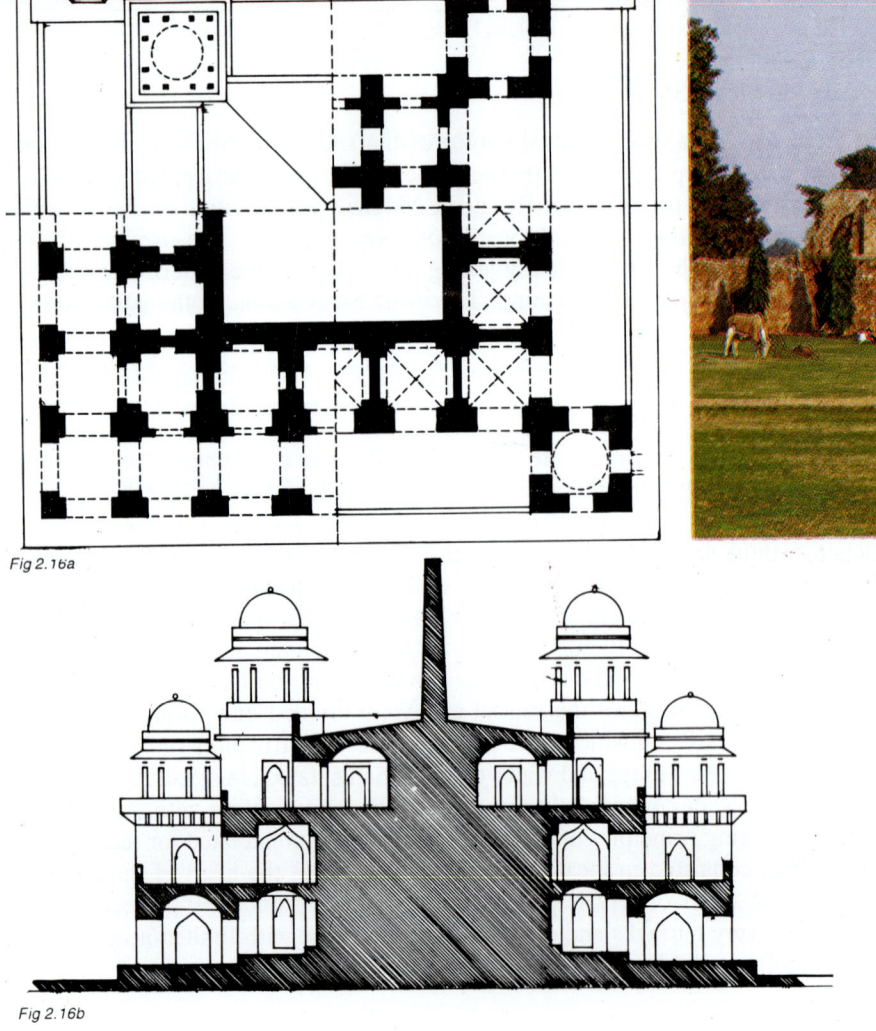

Fig 2.16a

Fig 2.16b

Fig 2.16 *Three-tiered structure at Kotla, Delhi. (a) Plan, (b) Section*

Fig 2.17 *View of the Ashok lat and the substructure at Kotla, Delhi*

by building sloping, buttressed walls for the entire structure, but by locating tapering circular planned quoins at the corners and in the rear of the *maqsura* and at the entrance points. A more prominent symbol of the pronounced militarism of the architecture was a massive arched and buttressed pylon-like structure invariably planted in the middle of the *maqsura* facade. This central feature was so tall that the huge dome over the central compartment of the *liwan* was not visible from the courtyard. It would seem that the builders found the straightforward three-arched facade of the Jamat Khana mosque rather too insipid. To them, the artificially planted screen of arches at the Quwwat-ul-Islam had a flamboyance lacking in the 'honest' facade of the Jamat Khana. Though the Tughlaq builders built their *liwans* over bays composed of stone columns and 'tudor' arches, the ceremonial pylon, not different in intention to the 'screen of arches', became a popular innovation. The most striking example of this is in the Begumpuri Masjid near modern Malviya Nagar *(Figs 2.18a, b)*.

Another design device that added to the grandeur of the Tughlaq mosques was that the courtyard of these was generally built over a platform or basement often raised more than 12 ft (3.6 m) above ground level. This necessitated the building

Fig 2.18 Begumpuri Masjid, Delhi, AD 1370. (a) The entrance gateway, (b) Courtyard

Fig 2.18a

Fig 2.19 View of the raised entrance of Kalan Masjid, Delhi, AD 1375

Fig 2.18b

of imposing flights of steps leading from ground level up to the entrance gateways as in the Kalan Masjid (*Fig 2.19*). The design of the entrance gateways, too echoed the form of the central pylon dominating the *sahn* or courtyard. The lower periphery of the erected basement became deep arched niches, generous enough in size to be put to use either as living rooms for the attendant priests, or as shops, or even as dormitories for pilgrims on festive occasions.

The Covered Court of the Khirki Masjid

All the mosques except the so-called Khirki Masjid of the Tughlaq period fall into this general pattern varying only in size and details. In the Khirki Masjid, on the contrary, the builders attempted a mosque of a type hitherto unknown to India. Obviously, the scorching sun of the long summers of India was the chief motivating force behind this new design. In planning, a part of the *sahn* was covered by a combination of a domed and flat roof, leaving four symmetrically arranged open-to-sky courtyards for light and

Fig 2.20 Khirki Masjid, Delhi, AD 1375. (a) External facade, (b) Bird's eye view, (c) Open court, (d) View of bays

Fig 2.20a

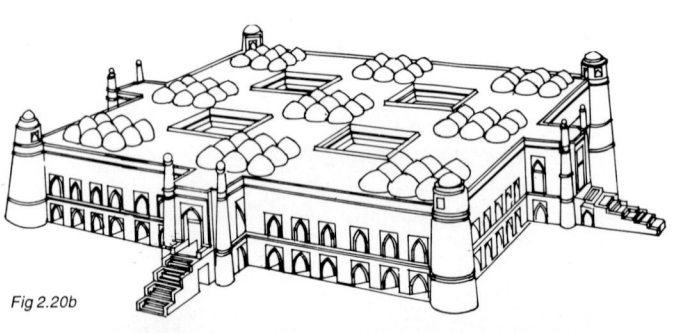

Fig 2.20b

Fig 2.20c

Fig 2.20d

ventilation (*Figs 2.20a, b, c, d*). No doubt the hot Indian sun was cut out, and more comfortable praying conditions created. Unfortunately, this device divided the space of the open courtyard into definable small spaces. To the great 'egalitarian brotherhood' of Islam where congregational worship was the pivotal ritual, this 'compartmentalization' of the faithful at prayer was psychologically more difficult to tolerate than the hot Indian sun. And so this experiment was repeated only once again at Gulbarga in the south. The fact of the matter is that the 'congregational' aspect of prayers is inviolate for the Muslims. The choice was either to be able to build the uninterrupted huge domed spaces of the Turkish mosques (where snow in the winter prevented the holding of prayers in the open), or be content with open courtyards. The latter remained the dominant element in the design of the Indian mosque.

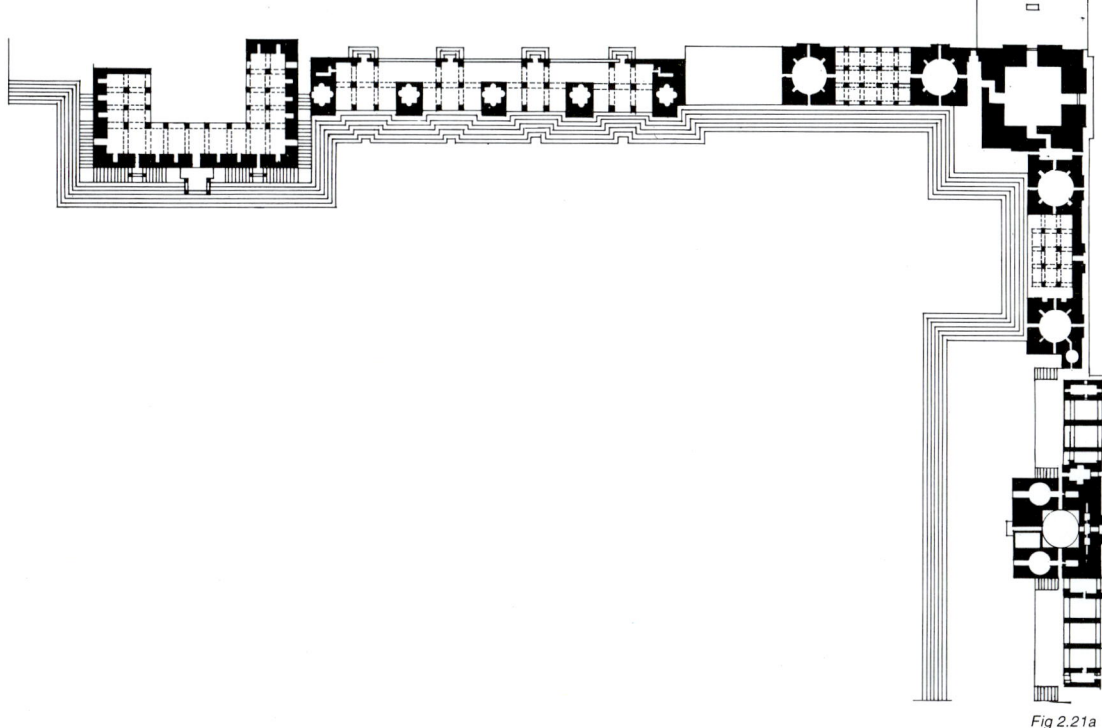

Fig 2.21a

Firuz Shah and the Hauz Khas

The Indian tomb under the innovative hands of Firuz Shah's builders followed the formal aspects set by Ghias-ud-din, though built with inferior materials. But Firuz Shah, with an uncanny eye for situation rather than mere self exaltation, decided to be buried in the unostentatious but beautiful environment of the Hauz Khas built by Ala-ud-din Khalji some seventy-five years earlier. The Hauz Khas palaces had obviously remained a favourite picnic spot for the Khalji kings. Comprehending the innate peace and beauty of the surroundings, and in order to fulfil his pious ambitions, Firuz Shah decided to build a mosque at the northern end of the existing tank and to install a *madrassa* or 'college of theology' in the buildings along the northern and western banks *(Figs 2.21)*. The college buildings are two-storeyed domed and

Fig 2.21 The buildings around Hauz Khas. (a) Plan, (b) View

Fig 2.21b

pillared halls. The upper storey is at ground level with ample cross-ventilation, while the lower is just above water level and closed on one side. The two storeys together were an ideal combination to combat the varying extreme climates of Delhi. At the corners where the two wings of the college buildings met, Firuz Shah decided to erect his own tomb. The tomb is a beautifully proportioned 45 ft (13.7 m) square structure, built in the characteristic ascetic style of the Tughlaqs. The familiar rubble masonry walls are finished with a thick layer of lime plaster punctured with arch and lintel openings, the whole crowned with a parapet of merlons. The handsomely contoured dome appears to rise over a base of trilobed merlons. The interior of the tomb is finely decorated with a geometrical design cut into thick layers of plaster rather than in stone, obviously for economic reasons.

Fig 2.22 Khan-e-Jahan Telengani near Nizammuddin, Delhi, AD 1368–69. (a) Plan, (b) View of the existing ruins

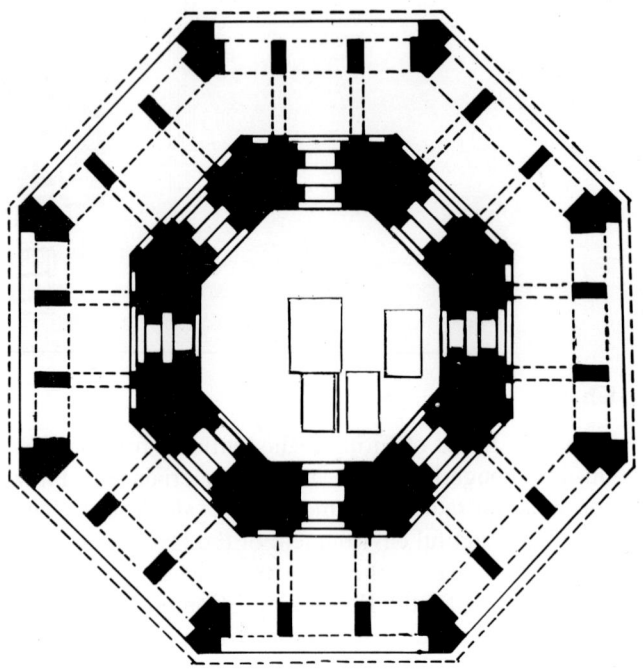

Fig 2.22a

Fig 2.22b

The Octagonal Tomb of Telengani

Structural economy once again must have been at least one of the reasons for the development of an altogether different type of tomb that appeared at the end of Firuz Shah's rule. This was the tomb of Firuz Shah's Prime Minister, Khan-e-Jahan Telengani, built in the Nizam-ud-din Auliya area *(Figs 2.22a, b)*. As we have seen, all tombs built in India to date had been square in plan except for the underground crypt of the Sultan Ghari which was octagonal in plan. The Tughlaq builders decided that since the inner space of a tomb served only the function of accommodating a grave, it need not necessarily be square in plan but could very well be an octagon. Over this, the circular dome could be installed without going through the cumbersome structural process of arching across the right-angular corners of the square to arrive at the octagon and finally the circle for the dome. The builders may well have been inspired by the similar octagonal plan of the sacred Mosque of Oman in Jerusalem in surrounding their crypt with a veranda on all its eight sides. It may also be possible that the dome, sitting so transparent and squat over an octagonal cylinder would have appeared more like a fat and oversized minaret than a structure having a tomb. The spread-out base provided by the surrounding veranda certainly lent visual credence to the structure. The entire composition is further appropriately graded by the installation of small kiosks (from the Turkish kiosk) along the base of the dome and over the veranda. What is unmistakably an indigenous innovation is the introduction of the deep sloping *chajjas* that sail out over brackets above the triple arched facades of the veranda. The use of the typical Hindu *chajja* added a new dimension to the otherwise plain surfaces of Islamic architecture in India. Under the host Indian sun these had been not only the traditional 'sun-breakers' but could just as effectively be used as 'visual breakers' to create a play of light and shadow on the facades. This first octagonal tomb though rather ungainly in its proportions, proved successful enough to encourage builders to experiment with its architectural potential right up to the beginnings of Mughal rule in India some two hundred years later.

Timur Burns Delhi

The great but religiously bigoted Firuz Shah fortunately did not live long enough to see the plunder of his beautiful city of Delhi by his co-religionist Timur, the ferocious grandson of Changez Khan, the Mongol. The seeds of Timur's invasion were sown, it would seem by Firuz Shah himself whose weak rule in the last years had encouraged a short of civil war in Delhi. His death was followed by the usual conspiracies and treacheries. Ultimately, 'the government fell into anarchy, civil war raged everywhere, and a scene was exhibited, unheard of before, of two kings (Nusrat Shah and Mahmud Shah) in arms against each other, residing in the same capital. The warfare thus continued as if it were one battle between the two cities wherein thousands were sometimes killed in a day.' But, this was nothing compared to what was fated for the inhabitants of the city of Firuz Shah's Delhi. In December 1398, Timur, 'who all along his route (of invasion into India) created a wilderness adorned with pyramids of skulls of those he had slain,' defeated the armies of Mahmud Shah outside Delhi. Thereafter, on the filmsy excuse of having detected a 'spirit of resistance by the infidel inhabitants' he let loose his army of fifteen thousands ferocious Turks to slay, plunder and destroy at their will. The Hindus were forced to set fire to their own houses along with the women and children, and an incredible number of more than a hundred thousand of them were 'despatched to hell by the proselytizing sword.' Timur, having done enough to earn the title of the 'Scourge of God' then decided to return to his native land having fulfilled his twofold aim of *jehad* and plunder, taking back with him hundreds of building craftsmen and materials to embellish his own city of Samarkand. But Delhi's miseries were far from over. For 'at this time such a famine and pestilence fell upon Delhi that for two months not a bird moved a wing in Delhi.' Ultimately, Sultan Mahmud Shah returned to rule nominally. With his death in AD 1413, the dynasty founded by Ghias-ud-din Tughlaq came to a more too glorious end.

The Builders of Delhi Migrate

Meanwhile, the building craftsmen had few patrons left in Delhi. A good many of them had been carried away in Timur's 'baggage-train' to Samarkand. Those that were left had no choice but to look elsewhere for building commissions. With the dissipation of Tughlaq rule in Delhi, Muslim governors of erstwhile provinces like Bengal, Jaunpur, Gujarat and Malwa proclaimed their independence from the sovereignty of the Delhi Sultanate. They were now busy glorifying and Islamizing their seats of power. These independent rulers became the new clients of the building craftsmen. The architectural scene for some time then, until just before the arrival of the Mughals in Delhi, shifts to some of these regions that were humming with building activity.

*Jahaz Mahal,
Mandu*

CHAPTER 3

Feudalism in Central and East India

AD 1305–AD 1500

Both politically and architecturally the scene in the various provincial outposts of Islam shown in the accompanying map of India *(Fig 3.01)*, was in many ways a reflection of Delhi. In the earlier centuries of Islamic India, Delhi was the fountainhead as much of Islamic culture and political intrigue, as of architectural inspiration, at least for the eastern and central regions of Jaunpur, Mandu and Bengal. As such, it may be redundant to trace the history of the succession of kings, governors, ministers and usurpers that grasped and sustained power in these provinces as and when opportunity arose. Suffice it is to say that the conditions were similar to those that prevailed under the multitude of feudal Hindu kings that had held sway over these areas, except that the source of temporal and religious power was now Islamic instead of Hindu. The king and the mullahs were busy building up their respective spheres of civic and religious influence, though not necessarily in the conspiratorial style that the Hindus had sanctioned to the Kshatriya king and his

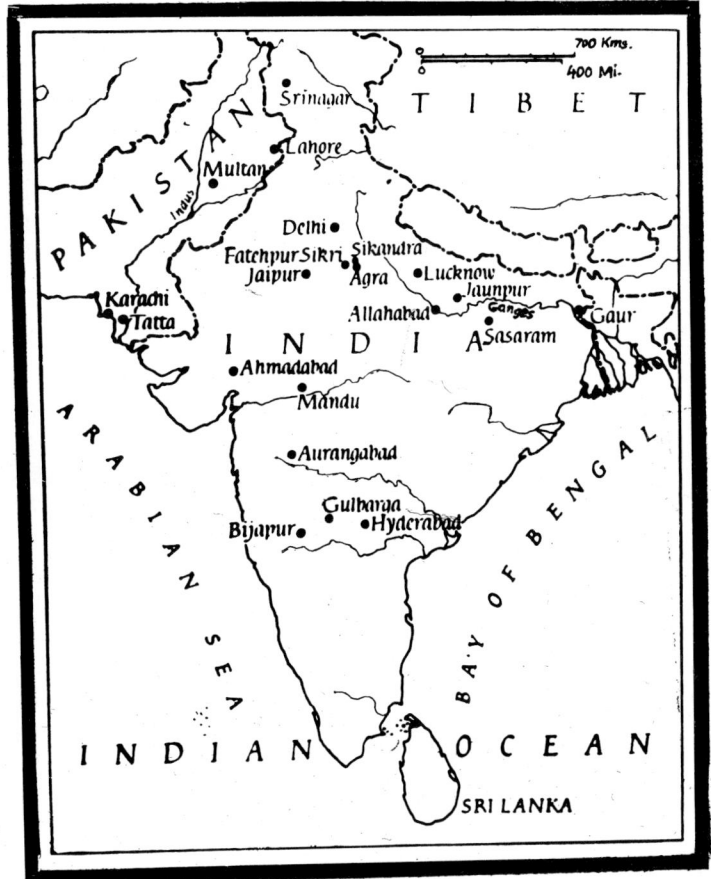

Brahmin ally. The attitude of Muslims towards monarchy was an undefined mixture of democracy and authoritarianism, depending on whether one was Shia or Sunni, close to or far from the Khalifat, a Turk, a Mongol, a Persian or a recent convert. The Muslim priest and ruler were more at loggerheads with each other than the Hindu Brahmin and his king were. This conflict was not sharp enough to discourage the king from financing the building of mosques and self-glorifying tombs. Only the number of mosques built was not quite as plentiful as temples had been — a phenomenon reflective of statistics, Mosques were required only for a 'minority' group since Islam, in spite of being the acknowledged State religion, was far from able to command universal following among the Indian people; the majority were too deeply submerged in the oceans of Hinduism. Moreover, because Islamic worship was congregational in nature, a single mosque served the same needs as a group of temples would. It was, therefore, politically imperative for the Muslim rulers that mosques, large and small, be quickly erected in the cause of proselytization and spread of the religion.

Makeshift Provincial Mosques

For the instant erection of mosques, the methods followed at Delhi were too conveniently attractive to be ignored. The earliest mosques in most of the provincial regions were thus no more than a recomposition of building materials extracted from existing Hindu and Jaina temples. However, only a few of them came up even to the rather minimal architectural standards of the Quwwat-ul-Islam that had been built in Delhi under similar circumstances. It would, therefore, not be worthwhile to describe in detail the 'reassembled' mosques of each region. In every province the initial efforts followed the general pattern established at Delhi. The differences in detail of the structural and decorative techniques of these makeshift mosques were governed as much by the quality of available 'Hindu' building materials and craftsmanship as the designer's ingenuity. Over a period of time, however, again as at Delhi, the provincial builders gradually created the so-called language of Islamic ideas. The most lively elements even of this 'pure' language, however, were those of indigenous craftsmanship, local building materials, and the climatological and social aspects of a province. So much so, that each region freely developed its own dialect of regional architecture that was often more expressive and fascinating than the more orthodox building vocabulary of the 'parent' family of Delhi. Inspiration from the architectural style of Delhi, though, was inevitable — at least for all the Muslim principalities that were within the geographical and wavering political ambit of Delhi.

Jaunpur and the Sharqi Dynasty

The most prominently distinctive style inspired by the architecture of Delhi was that practised in Jaunpur, an 'on-and-off' province of the Delhi Sultanate located just 36 miles (58 km) south-east of the Hindu holy city of Varanasi. Picturesquely located on a spur overlooking the banks of the river Gomti, the city of Jaunpur is said to have been established in AD 1360 by Firuz Shah, the last great Tughlaq emperor. According to Islamic legend he named the city after Jauna, his predecessor. It is more likely, though, that Jaunpur was the Muslim version of the ancient Hindu name of Yavanpur. However, it was the 'militaristic' aspects of the Tughlaq style that formed the kernel of the Islamic architectural style of Jaunpur. These were progressively muted into an acceptable 'civil' style in Jaunpur under the rule of the Sharqi (literally, 'eastern') dynasty. This process of mutation can be traced through three prominent mosques of Jaunpur that are the only significant survivors of the past glory of the city that today is a mixture of stately but sadly neglected mosques, and a haphazard cluster of bazaars, humble huts and houses.

Lal Darwaza Masjid

The smallest of these mosques is the so-called Lal Darwaza Masjid *(Fig 3.02)* built during the reign of Mahmud Shah in the year AD 1447. The edifice may only partly be classified as an 'archaeological miscellany' since it carried within it the seeds of an originality that were later developed by the builders of the great Jami Masjid of Jaunpur. The most prominent feature of Tughlaq architecture that caught the eye of the Jaunpur builders was the militaristic, buttress-fringed central pylon of mosques as at the Begumpuri at Delhi. But in its appearance in the Lal Darwaza Masjid at Jaunpur, muted as it was by the appurtenances of Hindu architectural features, the propylon was no longer reminiscent of fortress architecture. Rather, it is more like a crystallized version of the corbelled 'screen of arches' of the Quwwat-ul-Islam at Delhi.

Fig 3.02

Fig 3.03

Fig 3.02 Lal Darwaza Masjid, Jaunpur, AD 1447

Fig 3.03 Flat-roofed Hindu pillared cloisters and verandas, Lal Darwaza Masjid, Jaunpur

Even when the stock of readymade columns and beams from Hindu temples was exhausted, new ones on the old pattern but without the objectionable anthropomorphic representations of gods and goddesses were manufactured and assembled to form traditional flat-roofed cloisters and verandas *(Fig 3.03)*. It was almost as if the building operations were neatly compartmentalized. While workers familiar with the Muslim style were appointed to carry out the more prominent and sensitive components such as the *liwan*, local Hindu craftsmen were simultaneously allowed to put together the necessary ancillaries such as surrounding cloisters and facades. The rulers seemed to be keen to assemble their monuments quickly together. It was almost as if they knew that their Sharqi dynasty was not destined to last longer than a century. In this hurry the frank and honest juxtaposition of the two styles — one trabeate and the other clearly ornate — lent a distinctive and refreshing flavour to the Muslim architecture of Jaunpur.

The 'Ladies Only' Chambers

The builders of Jaunpur dealt effectively also with a sociological aspect of Muslim worship — that of the ladies of the court taking part in the ritual of worship, albeit in *purdah* (literally, 'in curtain'). Translated into building, this *purdah* became an elevated platform on either side of the *mimbar* within the *liwan* of the mosque, duly screened off by panels of *jaali*. From these enclosures the ladies could participate in the rituals of the *masjid* without being seen by the general assembly of the faithful. In the particular case of the Lal Darwaza Masjid, though, this provision for the fairer sex was probably ordinated by Bibi Raja, the queen of Mahmud Shah, who had the mosque built probably as a private chapel attached to her palace. However, the 'ladies only' enclosure remained a persistent feature even of the public mosques, the court of Jaunpur being romantic and civilized enough to indulge the religious inclinations of its ladies without exposing them to the gaze of the male worshipper.

The Lal Darwaza Masjid measured about 160 ft (48.7 m) square in rather an effeminate and delicate structure with its central pylon a modest 49 ft (15 m) high, and the otherwise militant buttresses subdued to form a flat and lacklustre central composition. Its predecessor, the so-called Atala Masjid, however, displays the more flamboyant aspect of the Jaunpur style. The popular name of this mosque derives from a temple to the goddess Atala Devi, which stood on the site and was destroyed to make way for the *masjid* to be erected over its foundations.

Fig 3.04 One of the central gateways, Atala Masjid, Jaunpur, AD 1408

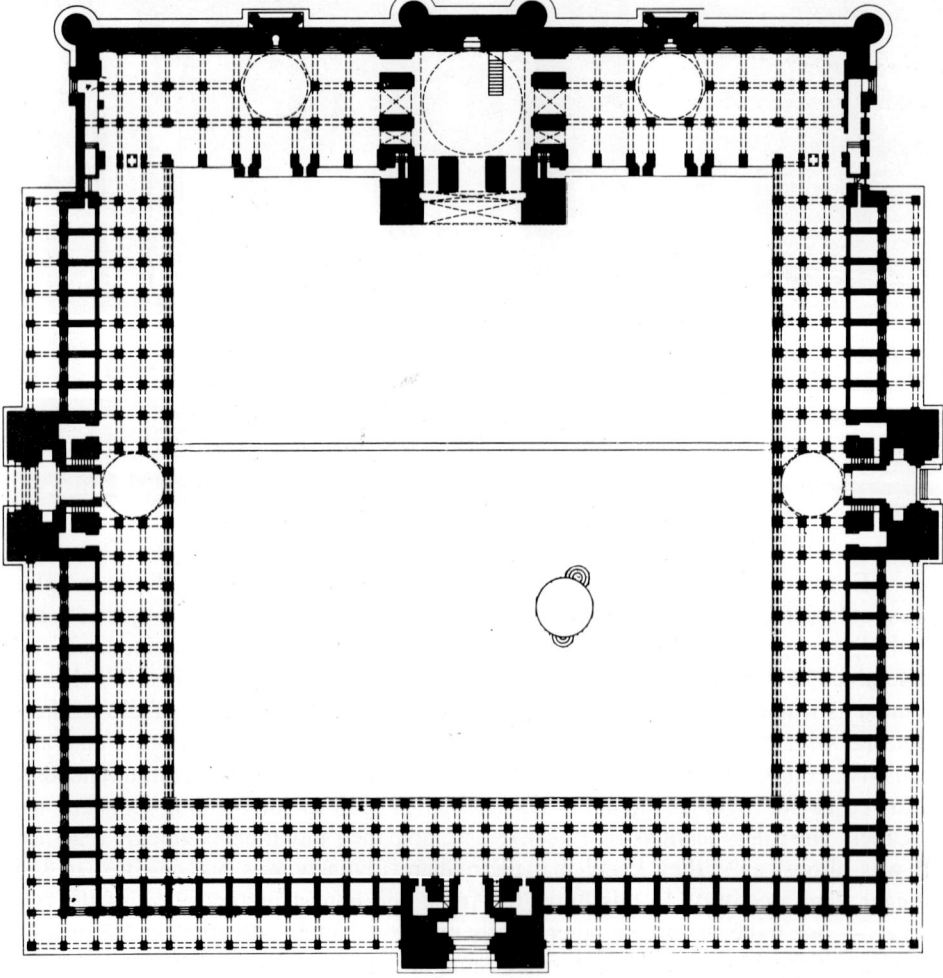

Fig 3.05 Plan of the Atala Masjid, Jaunpur

The Pulsating Rhythms of Atala Masjid

One of the earliest mosques to be erected at Jaunpur, the Atala admittedly displays the influence of the Delhi Tughlaq style of architecture, but has an additional flavour and vigour all its own *(Fig 3.04)*. This is most apparent in the robust design of the inevitable pylon in the centre of the Atala *liwan*. The circular tapering shafts of the Tughlaq model are now resolved into rectangular turrets while retaining the inclined profile of the original. Suspended between the two rectangular turrets is a huge spandrelled arch which, too, is more purposeful and majestic in its firm outlines than the wavering and ornamental ogee arch of the Tughlaqs. The typical Hindu bracketed openings as usual find their own place at the bottom of the arch, its upper reaches being filled in with arched apertures, *jharokhas* and *jaalis*. The builders of the Atala Masjid seemed to have been fully aware of the dramatic quality of the pylon that they had devised. Quite appropriately they not only installed two identical mini-pylons on either side of the central one, but also designed the three gateways in the centres of the eastern, northern and southern colonnades of the courtyard to echo the style of the *maqsura* pylon *(Fig 3.05)*. The silhouette of these various pylons of different sizes in the mosque sets up a pulsating rhythm that is not felt in any other mosque in India.

It is probably this same feel for rhythm that prompted the Jaunpur builders to infuse a liveliness even into the composition of the rear wall of the *liwan*, an element that was traditionally left as a plain pile of masonry in most mosques. Thus, the rear wall of the Atla Masjid is adorned with central and side projections, each related in intent and size to the domes that crown the rectangular hall of sanctuary. However, for some obscure reasons, the idiom used to define the ends of the projections is the same old Tughlaqian one of tapered circular buttresses rather than the rectangular ones that adorned the front of the Jaunpur *liwan*. This almost casual neglect of harmony strikes the only discordant note in an otherwise vibrant little gem of Islamic architecture in India. Inscriptional evidence testifies to the fact that the Atala Masjid was the work of a Hindu architect. 'That Hindu artisans were largely employed upon the works' is only indicative of the fact that 'freed from their age-old indigenous conventions' they were more than capable of 'inventive formation and infusing fresh spirit into such a notable architectural synthesis.'

Jaunpur's Jami Masjid

When the structurally more adventurous Muslim builders of Jaunpur set forth to build their next great mosque, they certainly succeeded in their grand constructional concepts but fared none too well in achieving equally great architectural effects. This is apparent from the form and shape of the great Jami Masjid of Jaunpur that look over a generation (from AD 1438–1478) to complete. Considering the intent and ambition of the builders, this was probably a none too long period. Of this mosque, the square inner courtyard alone, which is 200 ft (61 m) wide is larger than the entire mosque of Atala. What is more, it was built over a platform that was raised to a height of 20 ft (6 m) above ground level. This necessitated the building

Fig 3.06 View of domed gateways and two-storeyed cloister, Jami Masjid, Jaunpur, AD 1438–78

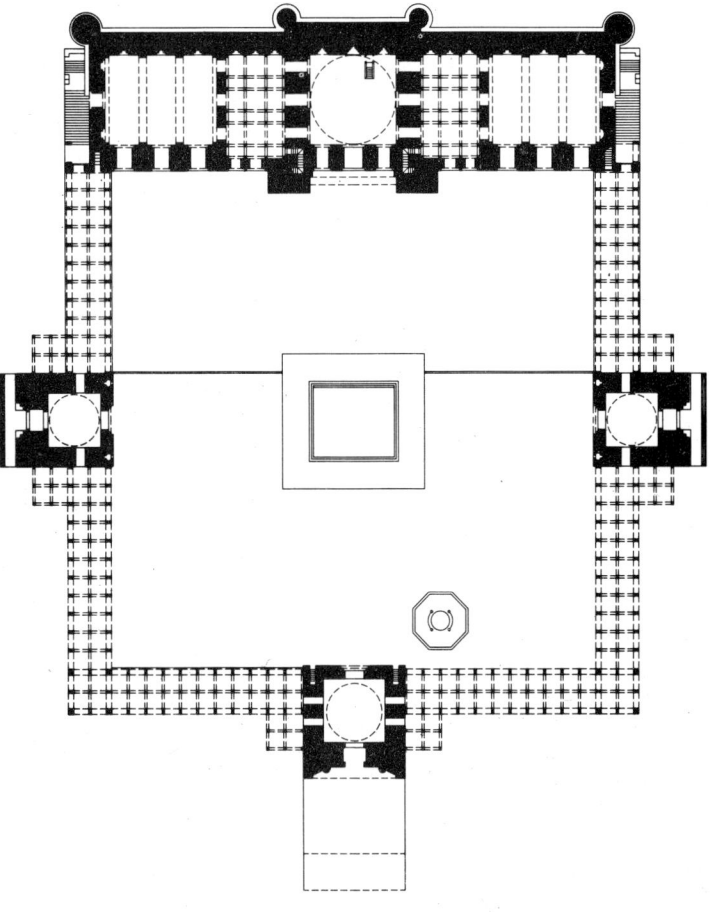

Fig 3.07 Plan of the Jami Masjid, Jaunpur

Fig 3.08 Isolated splendour of the pylon of Jami Masjid, Jaunpur

of imposing flights of steps, not unlike those of the Tughlaq mosques of Delhi, to reach platforms over which were erected lofty domed gateways leading to the courtyard. The *sahn* of the mosque was surrounded on three sides by an unusual two-storeyed cloister in keeping with the scale of the entire concept *(Fig 3.06)*. This was erected in the usual trabeate Hindu system, by imposing column over column, and was covered with a flat roof.

Vaulted Shells of the Liwan

It was in the construction of the prominent *liwan* side of the courtyard that the builders experimented with the structural form of the vault. The vault had been extensively used over a millennium ago in India by Buddhist carpenters and occasionally by brick masons in building their *chaitya* halls. The timid Hindu engineers, having discarded even the use of the arch, had never taken recourse to such daring measures. In a way, they had no need to, since the very concept of individual worship in the Hindu temple precluded the building of large open halls. However, for congregational worship in a mosque, pillarless halls were undoubtedly a desirable asset. Until now, the Muslim sanctuary or *liwan* had been spatially organized to consist of a central domed compartment flanked on its own southern sides either by smaller domed spaces or even colonnaded halls. The Jaunpur builders, in an attempt to create vast pillarless spaces on either side of the 40 ft (12 m) diameter span of the enclosure for the ladies, decided to roof over the halls with a longitudinal vault rising to a height of 45 ft (13.7 m). The vault was constructed over 'four pointed arches or ribs consisting of transverse ribs at wide intervals in the middle, and two wall ribs or formerts at each end. Over this permanent centering was erected a solid stone shell built of large blocks. The building of this shell was truly a commendable structural feat, equally awesome being the vast interior space so created *(Fig 3.07)*.

However, the load of such a vault was thrust entirely on the long eastern and western peripheries of the sanctuary, which inevitably became massive supporting walls, in this case as much as 10 ft (3 m) thick. What is more, openings in the critical eastern wall had to be restricted to arched apertures rather than generous doorways. The spatial link between the sanctuary and open courtyard was effectively diminished if not virtually snapped. This barrier between the *liwan* and the *sahn* was as intolerable to egalitarian Islam as had been compartmentalization of the covered courtyard in the Khirki Masjid at Delhi. Thus, like the example at Delhi, in spite of the inherent advantages of a spacious *liwan*, this experiment of a vaulted hall at Jaunpur was destined to remain an isolated one in Islamic architecture in India. Even the great propylon mounted in front of the *liwan* was built for the last time in this Jami Masjid. In spite of its grand Egyptian scale — rising as it does to a height of 85 ft (26 m) from a base of 75 ft (23 m) — without its mini counterparts on either side, it was reduced merely to a lofty gateway, rather than a viable component of a larger architectural scheme. In its isolated splendour, the Jami Masjid had lost its architectonic *raison d'etre (Fig 3.08)*.

Fall of the Sharqis

Whether the Sharqi builders would have repeated such a gargantuan adventure is a question that must remain unanswered since no further significant building venture was undertaken in Jaunpur after the Jami Masjid. The growing political ambitions of the last Sharqi ruler Hussain Shah were now directed at the capital city of Delhi, which in the sixteenth century was under the rule of the Lodis. This proved fateful not only for the Sharquis but for the city itself. After Hussain Shah was decisively defeated by Bulhul Lodi, subsequent retribution by the Lodis for Hussain Shah's temerity included the razing of virtually all the secular architecture of Jaunpur. And so, the city that under Ibrahim Shah Sharqi had acquired the reputation of the 'Shiraz of India' and the 'refuge of sages and literati', was stripped of all its architectural grandeur, except for its five sacred mosques.

Fig 3.09 Erected from the spoils of Hindu temples, Lat-ki-Masjid, Dhar, AD 1405

Twin Cities of Dhar and Mandu

Fortunately, such wanton destruction was not inflicted on another picturesque Muslim output — that of the region of Malwa — of which Dhar an Mandu were the twin capital cities. This region had been annexed from its famous Hindu Parmar rulers by Sultan Ala-ud-din Khalji in AD 1305. Timur's sack of the city of Delhi was the signal for the governors of Malwa to declare independence from the authority of Delhi, one of whose last Sultans, Mahmud Tughlaq had, in fact, sought refuge with the Malwa kings.

Initially, the Muslim rulers adopted Dhar as their capital city and assembled together the Kamal Maula and Lat-ki-Masjid by improvising with building material looted from local temples. The two mosques are in no way distinctive from other mosques of this category, except that they are more generous in size, and the central aisle of the latter is crowned with a handsomely contoured dome rising over a merlon fringed base *(Fig 3.09)*. As soon as Mahmud Tughlaq returned to Delhi after Timur's departure from India in AD 1401, one Alp Khan Ghuri assumed the paraphernalia of royalty and laid the foundations of one of the most grand and dramatic forts of Islamic India at Mandu.

Fig 3.10 Fortress walls of Mandu enclosing the vast area of the Vindhyan range, 18th century

Mandu, the Resolutely Stable

The Muslim rulers' preference for Mandu to Dhar as a capital city was initially prompted by reasons of security. 'The situation that Mandu presented of a natural barbican in the shape of a spur projecting from the Vindhyan range was ideal for such a purpose.' The plateau formed was at an elevation of 2,000 ft (610 m) above sea level, and was separated from mainland Malwa by the Kakra Koh, a winding gorge 300 to 400 yards (100 to 130 m) across, and more than 200 ft (61 m) deep. The builders enclosed an area of approximately 12 sq m (31 sq m) within fortress walls over 25 miles (40 km) in circumference *(Fig 3.10)*. The elevated plateau, in fact, consisted of 'undulating tracts shaded by trees, dark pools nestling in the shallows, and larger lakes glistening in the sunshine with sloping swards.' If this was not enough inspiration for the builders to create great architecture, they could any time look down from the gargantuan walls upon the 'vast plains of the river Narbada, the delicate opalescent tints of which, formed by distant cultivation and winding waterways, formed an entrancing background.' It can be said without hesitation that the patrons and designers of Mandu did not fail their environment. Displaying a fine appreciation of nature, form, colour and materials, and the architectural intent of the various phases of their building activity being crystal clear, they succeeded eminently in enhancing rather than marring the rare beauty of the landscape entrusted to them, to achieve a commendable marriage of nature and man-made art.

The 'Swinging Mahal'

The first phase of serious architectural activity in Malwa, apart from the period of the makeshift mosque, was heralded by the construction of the so-called Hindola Mahal or 'Swinging Palace' constructed by the same Alp Khan who had built the city wall, and had now assumed the title of Hushang Shah. In its architectural manifestation, this phase is best described as being 'most decidedly stern and resolutely stable.' And the Hindola Mahal *(Figs 3.11, 3.12)* in spite of the illusion it creates of swaying like a swing is certainly most 'resolutely stable' — in fact, far more stable than structurally required for this modest sized building. It is indeed curious that both its 'swing' and 'stability' arise from the same cause: the six massive, sloped buttresses with deeply recessed arches that were erected to support the roof of a 60 ft (18.2 m) wide and 118 ft (36 m) long rectangular hall. This compartment is attached to another transverse hall of almost the same size at its northern end. The whole plan of the Mahal thus reads as a 'T' in shape. This earliest of Mandu structures, by its curiosity of design, has generated speculation as to its intent and function. The 'cross bar' and the 'stem' are both distinct in plan and concept — one a simple double-height rectangular hall, the other, two storeyed with only a gentle batter in its sloping walls, and its severe form diluted by typical Hindu balconies and *jharokhas*. It is most likely that the building served the purpose of a durbar or audience hall, the upright stem representing the public hall which appears to have been built first, while the cross bar, indicating the transverse portion by its two-storeyed configuration, suggests that it was a royal apartment. From the upper storey of this part the king gave audience to the populace gathered in the 'swinging' part of the palace below *(Fig 3.13)*. The two compartments may well have been erected at different times, but it is fairly reasonable to assume that its ultimate function was to serve as a 'royal apartment-cum-audience hall' for which in many ways, even the fortuitous additions seem ideally planned.

Fig 3.11 Jaali work in Hindola Mahal, Mandu, AD 1425

Fig 3.12 External view of Hindola Mahal, Mandu

Fig 3.13 View of upper and lower parts of Hindola Mahal

Fig 3.12

Fig 3.13

Fig 3.14 The stately domes of the Jami Masjid of Mandu, AD 1440

Solemn Silence of the Jami Masjid

Hushang Shah's Jami Masjid, on the contrary, shows no such signs of fortuitous planning. In fact, the very concept, design and subsequent construction of this *masjid* even after the death of Hushang Shah, are all part of a grand and dignified concept finished to near perfection. Very few mosques in India, including those of the Great Mughals, can touch this one in its grand and unique achievement of reconciling the abstract concept of the Islamic mosque with the most humane qualities. Even as it stands today, stripped of its gentler artefacts such as *chajjas* and characteristic coloured stone decorations, the very air of its stately courtyard strikes one with awe. It almost appears as if the designers of the mosque had instilled an air of vitality into its open courtyard as intensely as the Buddhist builders had filtered glimmers of light into the dark caverns of Karli. What the magnificent Buddhist builders had achieved through blurring out of detail in the soft effulgence of light, the Mandu builders achieved by exposing minimal detail and volumetric form in the harsh sunlight. The vibrancy of architectonics in this mosque is generated through a clear and honest delineation of a few chosen Islamic architectural forms. The courtyard is surrounded on three sides by a uniformly proportioned gallery of majestic arches, with just a hint of the ogee at the apex *(Fig 3.15)*. Above these arches are planted cylindrical cupolas of a distinctively masculine order. These domes rise vertically from their circular base to converge near the apex, in profile not unlike the so-called shoulder-shaped contours of the *shikhara* of Orissa temples. A pulsating rhythm is infused into the regimented skyline by large domes of the same stately profile planted at the corners of the mosque *(Fig 3.14)*. Crowning the skyline is a massive and strikingly handsome dome over the middle of the *liwan*, making a total of 158 domes for this mosque. Considering, however, that the overall size of the mosque is a grand 288 ft (87.7 m) square enclosing a generous square courtyard of 162 ft (49.3 m) side, the domes are not too many.

Fig 3.15

Fig 3.15 Gallery of uniformly proportioned arches topped by cylindrical cupolas

Fig 3.16 Solemn interior of the Jami Masjid, Mandu

Fig 3.16

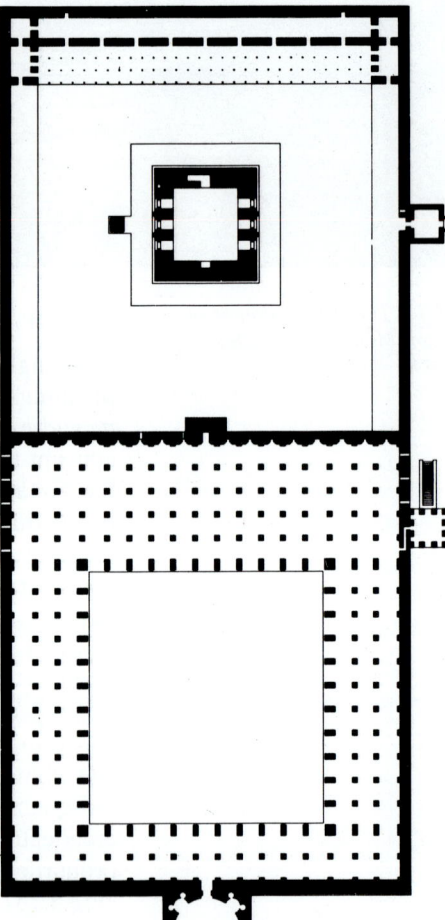

The interior of the *liwan* and other features of the Jami Masjid also adequately reflect its sacred purpose in being an 'assemblage of solemn silences, of muted passage with only minimum articulation' *(Fig 3.16)*. This controlled articulation amounts to austere screened platforms for the ladies on the *liwan*, broad stately flights of steps reaching up to the *sahn* platform, a domed square gateway, and arched niches in the basement of the platform to serve as a *sarai*. The bare structural bones consisting of square unadorned shafts for pillars, handsome arches and cylindrical domes, all speak more eloquently than many other fussed and fretted mosques of the Islamic world.

Fig 3.17 Plan of Hushang Shah's tomb and the Jami Masjid, Mandu, AD 1440

Tomb of Hushang Shah

Hushang Shah Ghuri's restless spirit led him into many a rash military venture and he had perforce to learn to taste defeat. He attempted, however, to turn his ultimate defeat into a triumph in marble, by erecting a stately mausoleum as his final resting place. In the fitness of things, Hushang Shah was laid to rest just behind the western wall of his great Jami Masjid. Built entirely in white marble *(Figs 3.17, 3.18)* his square tomb, though, is not quite as successful in its proportions as the Jami Masjid. The massive cylindrical dome seems too large for its square and squat base. This was probably due to the fact that the edifice was completed by his successor, Mahmud Khalji, who may have been in a hurry to conclude the construction without great concern for architectural niceties. Nevertheless, it is to the credit of the builders of Mandu that when Shah Jahan, the Mughal emperor, decided to build the Taj, he sent his architects to study Hushang Shah's mausoleum as part of their preliminary research. It is probable that the idea of planting four small domes around the central one in the Taj Mahal was inspired by Hushang Shah's tomb, where this Hindu concept of the *panchatayana* planning had been employed for the first time in Islamic architecture.

Fig 3.18 Hushang Shah's tomb, Mandu

Haft Manzil and Ashrafi Mahal

Sultan Mahmud Khalji's early life was largely spent on the battlefield forming alliances with other Muslim rulers in a bid to crush the famous contemporary Rajput chieftain Rana Kumbha of Mewar. During this early period when Mahmud's battle tent became his home,' building activity at 'home' seems to have languished. It was only when the Muslim forces scored a doubtful and short-lived victory over the Rajput forces that Mahmud decided to adorn his capital city of Mandu by building the celebrated Haft Manzil or seven-storeyed victory tower, of which only the base

Fig 3.19 Base of the Haft Manzil or seven-storeyed victory tower, Mandu

survives *(Fig 3.19)*. Constructed of the typical Mandu sandstone, it was said to have risen to a height of 150 ft (45.6 m), with each of its seven storeys demarcated with bands of marble. If one were to go by the quality of building achievements, Rana Kumbha must have been the victor of these Rajput-Muslim battles. While only the stump of Khalji's tower survives, Rana Kumbha's famous nine-storeyed Kirti Stambha at Chittor erected earlier to celebrate his victory over the rulers of Mandu, stands today in its full glory, certifying that 'Rana retained in his service the better builders.' The building standards at Mandu certainly seem to have deteriorated. The Haft Manzil was not the only victim of these falling standards. Khalji's own tomb, the so-called Ashrafi Mahal, erected over a makeshift basement prepared by filling in the central courtyard of an existing *madrassa* may, at one time, have deserved its name of 'Hall of Golden Mohurs.' But now, it is as much a heap of ruins as the Haft Manzil which was planted at one of the corners of the same *madrassa* basement. Only the gateway of this mausoleum, sited directly opposite the gateway of the Jami Masjid, shows that though quality of construction had deteriorated, the Mandu builders' fancies had not abated. The building seems to have been 'sumptuously embellished as each wall was faced with white marble, and the doorways, windows and cornices were elegantly carved, while in certain places patters in choice stones were inlaid with friezes of blue and yellow glazes.'

The Jahaz Mahal

It was obviously in the leisure hours of his later years which he spent 'hearing the histories and memoirs of the courts of different kings of the earth,' that Mahmud Khalji gave impetus to the beginning of the second phase of building activity in Mandu, 'when the style was beginning to progress towards that lightly elegant and fanciful mode.' The first somewhat restrained step was the building of one of the most popular structures of Mandu, the so-called Jahaz Mahal (literally, the 'ship palace') *(Figs 3.20a, b)*. The building earned its name by being located between two beautiful water bodies, the Kaphur or 'Camphor' Talao, and the other, Munja Talao, just half a kilometre north of the Jami Masjid, and close to the Hindola Mahal built some forty years earlier. This east-west oriented 360 ft (110 m) long and almost 50 ft (15.2 m) wide structure consisted essentially of a series of compartments and corridors partly built over the waters of the Munja Talao, and a number of airy and fanciful open kiosks on the broad upper terrace. Altogether, the building character is more lively and light-heartedly entertaining than the stolid dignity of the earlier

Fig 3.20b

Fig 3.20 (a) Jahaz Mahal or the ship palace, Mandu, AD 1460, (b) View through the ornate ortel windows

Fig 3.21 Detail of baths, Jahaz Mahal

Fig 3.20a

buildings of Mandu. That court life at Mandu was gradually sliding into one of 'pleasurable beguilement and carefree living' is fully expressed in the open-air baths on the terrace and swimming baths of fanciful design with triapsed ends and concave sides on the northern end *(Fig 3.21)*. Undoubtedly, the Khalji kings spent many hours in dalliance with the ladies of the court at these baths. At the southern end of the Jahaz Mahal are the remains of a complex system of waterworks that at one time have been equipped with devices such as Persian wheels to maintain a balance between the water tanks on either side of this 'ship palace.' A broad flight of steps, probably necessitated by the waterworks, takes one straight up to the terrace from the southern end, thereby throwing the deeply arcaded eastern elevation out of balance. At the same time it adds to the liveliness of this 'substructure with domed partitions on the terrace resembling a high floating hull with cabins on the deck and a captain's bridge overhanging the middle.'

Architectural Palimpsests at the Munja Talao

Seizing avidly on the potential generated by the Jahaz Mahal, the later Khalji kings surrounded the Munja Talao with a series of summer retreats and fancy palaces *(Figs 3.22a, b)*. These included the so-called Champa Baori which consists of ingeniously devised 'subterranean passages communicating with vaulted rooms disposed around the open shaft of a well, with one passage boring through on to the lake, at the edge of which was a pavilion affording a beautiful waterfront view of the lake.' Equally sumptuous must have been the Hawa Mahal (or Palace of Winds) on the western banks of the Munja Talao, where the designers certainly

Fig 3.22 (a) View of Manja Talao, Mandu, surrounded by summer retreats and palaces

went to town to please their indulgent masters. Fanciful oriel windows, merloned domes, pyramidal roofs, playful staircases and unprecedented circular shafted columns were all thrown in for good measure in building up an orgy of architectural palimpsests around the waterfront. In its heyday, despite its architectural waywardness, the *talao* and its waterfront architecture of summer houses, pavilions and palaces, must have provided an ideal environment 'away from reality where every whim and fancy of the rulers could be indulged in privacy or publicly practised.' It must have been this 'picnic garden' atmosphere of Mandu that attracted the romantic Mughal ruler Jehangir to visit Mandu repeatedly and spend a considerable amount on its upkeep and maintenance.

Having exhausted the space available on the banks of the Munja Talao, the successors of the Khaljis spread out their building operations over the exceedingly picturesque landscape within the city walls, and proceeded to erect an endless series of private pleasure pavilions. These generally consisted 'of a series of compartments around a central courtyard graced with pools and fountains, while above were arched loggias roofed with fluted domes, the surface everywhere gorgeous with painted tiles.' Included in this last and dying phase of Mandu, the City of Joy, are buildings reminiscent of the days of the great romantic legends associated with Baz Bahadur, the poet king, and his Hindu mistress, Rupmati. In spite of the fact that the sounds and laughter of this city of joy are no longer heard, and 'most of its monuments are scarred and blackened, the more imposing *durbars, hamams,* harems and reservoirs have been spared from time's dark fingers and still provide many enticements to both the lay and professional explorer.'

Fig 3.22 (b) The picturesque landscape seen through one of the private pleasure pavilions, Mandu

Bengal, the Explorer's Paradise

In fact, it was the sites of Muslim architecture in East India that were really an explorer's paradise, the region of Bengal being 'singularly inimical to the preservation of architectural remains. If the roots of a tree of the fig kind once find a resting place in any crevice of building its destruction is inevitable, and even without this the luxuriant growth of the jungle hides the building so completely that it is sometimes difficult to discover it... always to explore it.' Fortunately, a century after James Fergusson visited the sites of Bengal, the Archaeological Survey of India has done most of the more difficult explorations and restorations. Even the less adventurous can now discover for themselves the architecture of Bengal's capital cities of Gaur and Pandua, corresponding to Malwa's twin towns of Dhar and Mandu. But here, the similarity between Bengal and Malwa ends. No two Indian Islamic styles, in spite of both being prominently arcuate, could be so dissimilar. The reasons are more than that of the distance of 1,000 miles (1,600 km) that separated Bengal from Malwa. For one, in spite of the fact that Bengal was so far away from Delhi, it was one of the earliest provincial regions of India to capitulate to the powers of Islam — almost a hundred years earlier than Malwa. The answer to this anomaly lies as much in geographical as ancient historical factors. The great river Ganga afforded a direct means of communication right across the fertile plains of northern India. Just as the early Aryans had found it easy to traverse these vast plains to reach the abundantly cultivable areas of Bengal and the mineral wealth of Bihar, so did the Muslims in achieving their goal of capturing these two regions.

Tribeni and the Lack of Building Material

Having conquered Bengal in the beginning of the thirteenth century, the Muslim governors of the Slave Dynasty founded their earliest centre of power at Tribeni in the Hooghly district. Establishing political power and building mosques proved to be two entirely different propositions. The former was easier accomplished than the latter in the rain-deluged, flood-devastated and bamboo-forested plains of Bengal. What the early Muslim builders missed most here was an abundant supply of readymade building materials from extant temples. There were only a few places of Hindu worship built in stone, a material rarely available in the plains of Bengal. Those temples built in brick provided iconoclastic satisfaction to the demolishing squads. However, there was not sufficient building material for constructing their own mosques and tombs. Thus, the 'pure' reassembled mosque, entirely fabricated from looted beams and columns, did not form the foundations of Islamic architecture in Bengal. Rather, Muslim structures even up to a very late stage of the development of the building art in Bengal remained a curious mixture of stocky basalt columns of Hindu temples, and a superstructure of arched brick vaults and domes. With brick as the chief building material available, the so-called 'arcuate' style was inevitable. Among the earliest examples of this type of improvised structure standing in its original form is the tomb of Zafar Ghazi Khan at Tribeni. The typical basalt columns planted in the interior of the building are supplemented with voluminous quantities of new brickwork walls with small pointed arched openings. The pointed arch which became typical of this style is the so-called 'drop arch,' the centre from where it is drawn being located well below the springing point.

Adina Masjid at Pandua

In a great and surprising burst of energy, as if to celebrate the self proclaimed freedom of Bengal from the sovereign Slave Dynasty of Delhi, Sultan Sikandar Shah decided to embellish his new capital of Pandua with a huge Jami Masjid that came to be known as the Adina Masjid. The entire edifice covers a rectangle of 507 ft × 285 ft (155 m × 87 m) and contains within it a huge courtyard measuring 400 ft × 154 ft (122 m × 47 m) *(Figs 3.23a, b)*. The layout of the courtyard in such a manner that the longer side of the court faced the west marked a distinct departure from the conventional pattern of square courtyards. It is unlikely that the inspiration came from as far away as Damascus, where the great eighth century mosque of the Umayyads follows a similar configuration. Rather, as can be seen, a layout of this nature makes for a larger

Fig 3.23 (a) The crumbled ruins of the northern liwan of the Adina Masjid, Pandua, Bengal, AD 1364

Fig 3.23 (b) Plan of Adina Masjid, Pandua, AD 1364

Fig 3.24 View of the basalt pillars used in three- and five-angled enclosures around the courtyard, Pandua

proportion of covered area in respect of the total area, in comparison to the conventional plan. Under the climatic conditions prevalent in Pandua, roofed halls were more comfortable to worship in than open courtyards. Statistically, the Adina Masjid is undoubtedly impressive. The three- and five aisled enclosures around the courtyard are supported over about 260 pillars of basalt *(Fig 3.24)*, and when complete, were roofed over with no less than 378 brick domes. The central part of the *liwan* was at one time a very impressive pointed vault over 70 ft (21.3 m) deep. It spanned over a distance of 34 ft (10.3 m) and was supported at either end on walls perforated with five arches each. The apex of the vault was at a height of over 50 ft (15.2 m) and is now in ruins. What must have been the *pièce de résistance*, the central vault, has collapsed; the roof of the northern portion of the *liwan* is crumbled ruins scattered over the ground, and less than half of its numerous domes are intact. Apart from the vastness of its courtyards which 'presents the appearance of the forum of some ancient classical city,' and the ambitious vaulted hall in the *liwan*, there is little in the building that could be classified as great or original architecture. To the severe eye of Cunningham, the '*masjid* is little better than a gigantic barn.' Subsequent builders of Bengal, however, seemed to have learned two vital constructional and aesthetic lessons from this massive building venture. One, that to combat the severe tropical conditions of Bengal, a mere assembly of arches and domes on the conventional pattern would not necessarily result durable construction. Two, that if brick were perforce going to be the chief building material, they had a lot to imbibe from local building traditions, rich in the art of terracotta decoration of building surfaces. The subsequent story of the architecture of Bengal, thus, centres around the techniques employed by the Muslim builders to create architectural forms in brick that could subsequently manifest the intentions of Islam. Gradually, there emerged a building style that is worthy of description if not adulation.

The Eklakhi Tomb

The earliest monument of Bengal that can be said to have seeds of originality was the so-called Eklakhi (literally, 'one lakh') tomb at Pandua *(Figs 3.25a, b)*. For the first time, the Muslim builders derived their inspiration from the immediate environment and triumphed in their battle with inimical climatic conditions. To combat the incessant

Fig 3.25a

Fig 3.25 Eklakhi Tomb, Pandua, AD 1430. (a) Plan, (b) View

Fig 3.25b

and virtually unending rain of Bengal, the traditional Bengal villager built his roof with bamboo and thatch, the flexible quality of the former producing the typical camber of the Bengali roof. The builders of the Eklakhi tomb, taking their cue from the folk architecture of Bengal, duly modified the rigid 'cube and hemisphere' of the traditional Muslim tomb to one in which the flat portion of the roof was given discernible slope to throw off the rainwater. This marked slope of the roof, expressed rather elegantly in the bow-like formation of the parapets, became the outstanding feature of the Islamic architecture of Bengal. The surfaces of the brick masonry below the parapets of the tomb consist of a mixture of exposed brick fixed in bands to define door jambs and architraves, as well as in panels of cut and embossed brick depicting blind windows. Planted one over the other in tiers, these panels give to the building from outside the appearance of a three-storeyed structure *(Fig 3.25b)*.

The entire surface treatment of the Eklakhi tomb is, in fact, suggestive of the framework of the traditional wood and wattle hut, in spite of the entrances to the tomb being marked by stone gateways "torn bodily from a Hindu temple.' The rather well-preserved state of the building is clearly indicative of the fact that the absorption of the local idiom had not been in vain. The entire structural concept of this 75 ft (22.8 m) square mausoleum seems unduly cautious. The interior space between the square of 75 ft (22.8 m) side is a massive concentric octagon of masonry to support the rather modest plain hemispherical dome of the same diameter. To fill in with so much brick masonry was obviously a stratagem used to avoid the rather arduous task of converting the square of the plan to an octagon. The builders were not experienced enough to employ the cumbersome arched or pendentive squinches that would have then been required. Even though for political reasons in 1442 Haji Iliash, one of the most illustrious of Bengal's rulers, transferred the capital once again to Gaur (the traditional Lakhnauti) situated on a narrow strip of land between the Ganga and Mahanadi rivers, the architectural style of the Eklakhi tomb prevailed in the structuring of tombs, mosques, minarets or gateways.

The City of Gaur

Although the Archaeological Survey of India has certainly made innumerable structures of the ancient city of Gaur visible, only a few of these can today be read as monuments. These are the Dakhil Darwaza (AD 1465) *(Fig 3.26)*, the Gumant Masjid (AD 1484), Feroze Minar (AD 1488) *(Fig 3.27)*, Chhota Sona Masjid (AD 1510), Bara Sona Masjid (AD 1526). The Dakhil Darwaza gateway built by Barbak Shah is the most triumphant artistic achievement of the bricklayers of Bengal. This 75 ft (22.8 m) wide and 60 ft (18.2 m) high structure comprises a central vaulted or domed passage 13 ft (3.9 m) wide, with a series of guard rooms on either side of it *(Fig 3.26)*. The entrance facade of this 'saluting gateway' is evocative of the Delhi Tughlaq style, consisting as it does of two circular tapering turrets planted on either side of a lofty pointed archway. Similar to the pylons of the Tughlaq and Jaunpur mosque, these are enriched 'by a certain amount of ornamentation in terracotta, consisting of such motifs as flaming suns, rosettes, hanging lamps, fretted borders, decorative niches, and other patterns judiciously distributed.' The latter was the regional form of ornamentation that the Bengal builders applied successfully to the rather bare and militant appearance of Tughlaq architecture.

Fig 3.27 View of five-storeyed Feroze Minar

Fig 3.26a

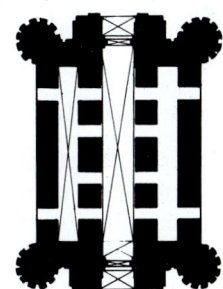

Fig 3.26 Dakhil Darwaza, Gaur, AD 1465. (a) View, (b) Plan

Fig 3.26b

Feroze Minar

Before getting immersed in the activity of producing mosques and tombs in the now well-established Bengal pattern, either Mahmud Khalji himself or one of his successors, inspired by the Qutb at Delhi, decided to adorn the city with a victory tower. The model for this modest 83 ft (25 m) high tower was not the 238 ft (72.5 m) high *minar* at Delhi, but one of the flanking towers of the Dakhil Darwaza at Gaur. Standing on a high artificial mound the tower has five storeys. The lower three are approximately 20 ft (6 m) in diameter having twelve sides, while the upper two are circular in plan. At the apex was installed a cupola over an open room that has now vanished. Vanished, too, are the blue glazed tiles that once decorated the now blank bands of terracotta, and gave to the tower its name of Feroze Minar or the Blue Tower.

Peculiarities of the Bengal Mosque

As for the mosques, the incessant rains of Bengal not only compelled the builders to change the mode of construction followed in building the Adina Masjid, but the very essentials of mosque design had to be changed. The mosque in Bengal could not be an open-to-sky courtyard surrounded by a colonnade. Rather, the substantial part of mosque had to be an enclosed hall, the courtyard being redundant during the long periods of rain. Thus, the so-called Gunmant Masjid on the outskirts of Gaur consisted of a hall 140 ft × 60 ft (42.6 m × 18.2 m) designed much like the Adina mosque *sans* the cloisters and the central vault *(Fig 3.28)*. The basic pattern of construction evolved in the Gunmant is that of building an outer peripheral wall of massive brick masonry (in this case, as much as 9 ft (2.7 m) thick), and then planting readymade square stone columns in the central portion. Over these were erected brick arches to create square bays that supported a series of domes. As mentioned earlier, the Bengal builders and Hindu temple building material in small quantities right through the 300 years of their building activity. As if aware of the limited supply of building material from Hindu temples, they either imposed a system of 'rationing' or carried out their plundering and building operations simultaneously. Thus, even in the later stages of building operation, temple columns seem to have been available to them for building their mosques, though they may have been well hidden behind facades of brick.

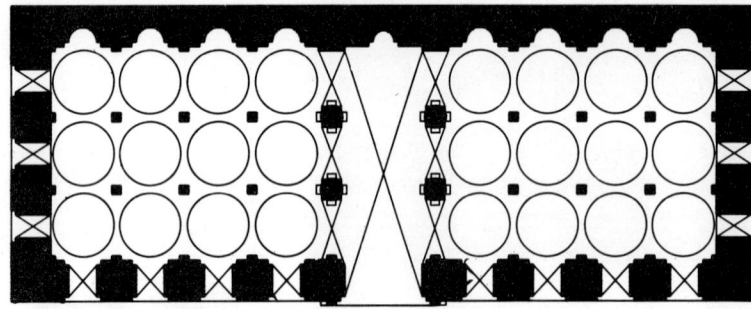

Fig 3.28 Plan of Gunmant Masjid, Gaur, AD 1484

From Chhota Sona Masjid to the Qadam Rasul

Hereafter, a series of tombs and mosques, all inspired by the style of the Eklakhi at Pandua, was erected at regular intervals in and around the city of Gaur. Apart from having acquired romantic names like Chor Khana and Lakhachippi in the romantic traditions of Bengal, the subsequent tombs are replicas of the Eklakhi and are more of archaeological than architectural interest. The mosques, too, once their concept had been reduced to that of a covered rectangular or square hall, differ from each other in terms of size, minor structural alignments and a flourish here or there. Thus, while the Chhota Sona Masjid *(Fig 3.29)* occupied a rectangle of 82 ft × 52.5 ft (25 m × 16 m),

Fig 3.29 Chhota Sona Masjid, Gaur, AD 1510. (a) View (b) Plan

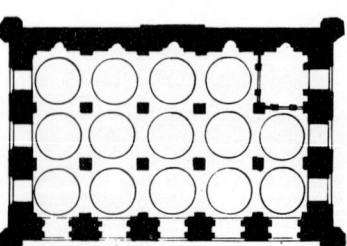

Fig 3.29a

Fig 3.29b

Fig 3.30

Fig 3.31

Fig 3.32

Fig 3.30 The typical Bengali 'drop arch'

Fig 3.31 Chamkatta Masjid, Gaur, 16th century

Fig 3.32 Tantipura Masjid, Gaur, 16th century

and its 15 cupolas were supported over eight columns and a mass of outer brick masonry perforated by the typical Bengali 'drop arch' *(Fig 3.30)*, the Bara Sona Masjid was a vast 176.5 ft × 162.5 ft (53.8 m × 49.5) in size with an inner *liwan* along its eastern periphery. Each of these mosques seems to have had gilt applied to its central Bengali roof-like cupola, which gave them their names of Small Golden Mosque and the Great Golden Mosque. Subsequently, while building the Chamkatta *(Fig 3.31)* and Lattan Masjid the supply of stone columns to support a series of small domes had run out, and hence the mosque was reduced to a mere large square compartment, much like a tomb with an attached portico. The whole Bengal movement ends on a rather feeble note in spite of the grandiose name for its last mosque, that of Qadam Rasul — literally, the 'foot of the Prophet' mosque. In plan *(Fig 3.33)* this 63 × 50 ft (19 m × 15 m) building consists of an inner compartment surrounded on three sides by a continuous corridor. The builders, having been able to lay their hands only on four basalt columns from Hindu temples, were forced to erect the more substantial part of their structure in brick. In the craftsmanship of brick, too, the vigour of the Eklakhi was missing. 'The charm and strangeness even of the curved cornice are handled in such a way as to lose their robustness, with the result that the whole structure tends to become flaccid and formless.'

The Bengal Roof

No wonder then that during the Mughal period, Fateh Khan, in building his own tomb near the Qadam Rasul, abandoned all the subtle techniques of his predecessors. Instead, he erected an imitation *in toto* of the Bengali hut complete with the curvilinear thatched roof, all built in brick and plaster. His effort proved to be not in vain. For, we shall see, it is the sensuous profile of this tomb that inspired the Mughals to incorporate the form of the Bengali roof *(Fig 3.34)* into the language of Mughal architecture, from where it found its true place of honour in the post-Mughal architecture of Sikh gurdwaras.

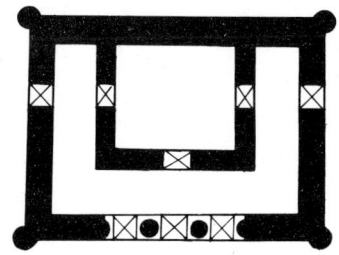

Fig 3.33 Plan of Qadam Rasul Masjid, Gaur, AD 1530

Fig 3.34 The Bengali roof form, Qadam Rasul Masjid

Pavilion at Sarkhej

CHAPTER 4

Ahmed Shah and Beghara of Gujarat

AD 1299–AD 1550

It was the Hindu architecture of the Gujarat region in India that completely enticed the provincial Muslim builders. This whole western sea coast region 'had been in a high state of civilization before its subjugation by the Mohammadans.' Ala-ud-din Khalji's armies wrested the reins of rule in AD 1299 from Karna Waghela, who ruled from his capital city of Karnavati. Once the seeds of Islam had been planted, in spite of valiant Rajput efforts the proliferation of Islamic ideals in the region was inevitable. The history of the region thereafter followed a familiar pattern: the gradual decline of Hindu power followed by the rule of Muslim governors appointed from Delhi, and the ultimate declaration of independence by the local Gujarat rulers from the suzerainty of the Tughlaq dynasty of Delhi. However, political independence from the authority at Delhi was easier gained than architectural freedom from the webs of the Gujarati builder. Attempts to do so were feeble and sporadic. The highly evolved Hindu art of the local guilds of building craftsmen prevailed for centuries, resisting Muslim rule. The tradition of these guilds of Gujarat, though ancient, was so vibrantly alive that the Muslim rulers had no choice but to 'appropriate to themselves almost *en bloc* the beautiful Gujarati style in the preparation of their mosques and tombs.'

Talents of the Gujarati Builder

The Muslim rulers of Gujarat became potentates, their treasuries being as rich as those of the earlier Jaina kings with the revenues from sea trade that fell under their purview. Yet it would seem that not a single one of the rulers was either able or willing to exercise personal influence over the style of Islamic architecture practised in his dominion. Thus, the Hindu or Jaina craftsman of Gujarat was left to deploy structural methods himself. He applied decoration and other architectonic techniques known to him to create a style that could be labelled as the 'Mohammadan architecture of Gujarat.' For this task the builders of Gujarat had been well-equipped by their ancient traditions of construction. Pre-eminent along these was a propensity for building the most lavish *mandapas* or front halls for their temples. Some of these *mandapas* had been fairly large halls roofed by a pyramidical corbelled dome held up over a ring of columns. The plan of such a *mandapa* could easily constitute a multipliable unit. A number of such conjoined units assembled together with minor design variations would easily produce large rectangular spaces or hypostyle halls necessary for the Liwans of the mosque. Moreover, the traditional Gujarati builder was capable of inducing dramatic spatial qualities into his temple *mandapas*, roofed with domes, cupolas and profusely carved horizontal ceiling panels situated at varying heights. In fact, the only new terminology the Gujarati architect needed to introduce to his vocabulary was that of the pointed arch, and the only he needed to delete was that of figurative sculpture. Once these two criteria were fulfilled his rather indulgent Muslim rulers were more than satisfied.

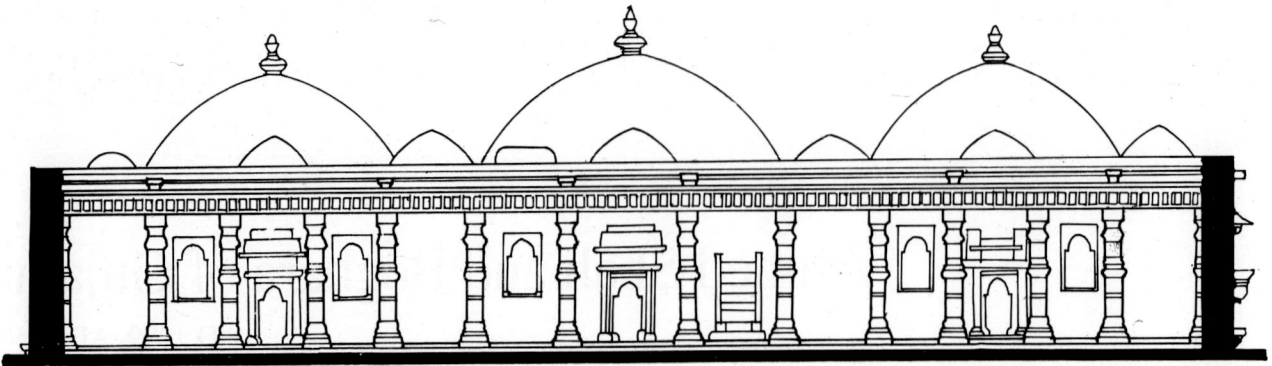

Fig 4.01 Elevation of Jami Masjid, Bharoach, 14th century

Jami Masjid at Bharoach

In the building of the earliest mosques in the region, however, there was little time to indulge in architectural niceties. Rather, there was time as usual only for downright plagiarism, some improvisation or, at best, a viable mixture of the two. The earliest Islamic building effort in Gujarat, begun soon after the raids of Ala-ud-din Khalji, was a clear case of improvisation. This was the Jami Masjid *(Fig 4.01)* in the ancient and prosperous seaport of Bharoach (ancient Bharakuchcha on the River Narmada). Columns recovered from the existing Hindu temples were replanted to form adequate octagonally aligned supports for erecting three corbelled domes of 30 ft (9 m) diameter, each separated from the other by a flat-roofed aisle and surrounded on all four sides by aisles of equal width. The long western walls and the two short sides were built up of solid masonry, with tiny arched and trellised apertures for cross ventilation, while the critical eastern wall, sheltered by a typical Hindu *chajja* below its parapet was left open. To the *liwan* so formed was added a courtyard defined by walls rather than verandas. However, as would be apparent from the elevation, even into this modest and rather bare structure, the Gujarati builder infused a stately quality by proportioning his front colonnade, *chajja* and domes in accordance with meticulous geometry borne out of centuries of building experience, strictly by the Hindu rules. The interior of the *liwan* with its rather ponderous bracketed columns may not have been ideally suited a mosque, but the facade is infused with an honest liveliness that was lacking in some of the more obviously Islamic oriented buildings in Gujarat.

Jami Masjid at Cambay

This quality of liveliness is particularly lacking in the totally plagiarized elevation of the Jami Masjid at Cambay (ancient Kambhat) erected some years later than the Bharoach example. The city of Kambhat, 'of which foreign merchants formed a chief part of its population,' was at one time under the rule of Vastupula, the famous Jaina minister of King Liwanprasada. Under this Hindu king, the town must have been studded with examples of Jaina architecture, some of which may have been as fine as the famous marble temples of Dilwara built under the same patronage. As an affront to the rich heritage of the Cambay builders, and as a mark of adulation to the Khaljis of Delhi, the new Muslim governors decided to emulate the Jamat Khana Masjid built at Delhi during Ala-ud-din's rule. The Gujarat craftsmen duly obliged. The results are as insipid as those of the Jamat Khana, insofar as the obviously plagiarized part of it is concerned. Thus, the facade of the *liwan* is composed of a blank screen wall consisting of a large central arch, and two symmetrically disposed subsidiary ones, with a merloned parapet *(Fig 4.02)*. The entire result, though admittedly Islamic in appearance, was rather like a dull blanket cast over the more pulsating rhythm of Hindu and Jaina columns and the fourteen domes that constitute the interior of the liwan behind. Fortunately, through the rather insipid screen of arches, a glimpse can be had of 'an engrailed arch of temple extraction, a motif which was afterwards to figure so prominently (in Gujarati Islamic architecture) as the graceful flying arch.' The cloisters that define the courtyard on the other three sides making for a mosque of 212 ft × 252 ft (64.6 m × 76.8 m) *(Fig 4.03)* are formed by a flat-roofed aisle and a series of twenty-one domes over octagonal column bays. The entrance gateway in the middle of the eastern cloisters is virtually a re-erected temple portico with a dome added on for Islamic effect. Just outside the southern cloisters exist the substructure of what was the tomb of Umer Bin Ahmed Kazaruni. The structure is akin to a grand Hindu or Jain temple *mandapa*, built up as it is by superimposing the columns to gain sufficient height. At one time it must have been crowned with the usual corbelled dome which has now vanished.

Fig 4.02 View of Jami Masjid, Cambay, AD 1325

Fig 4.03 Plan of Jami Masjid, Cambay

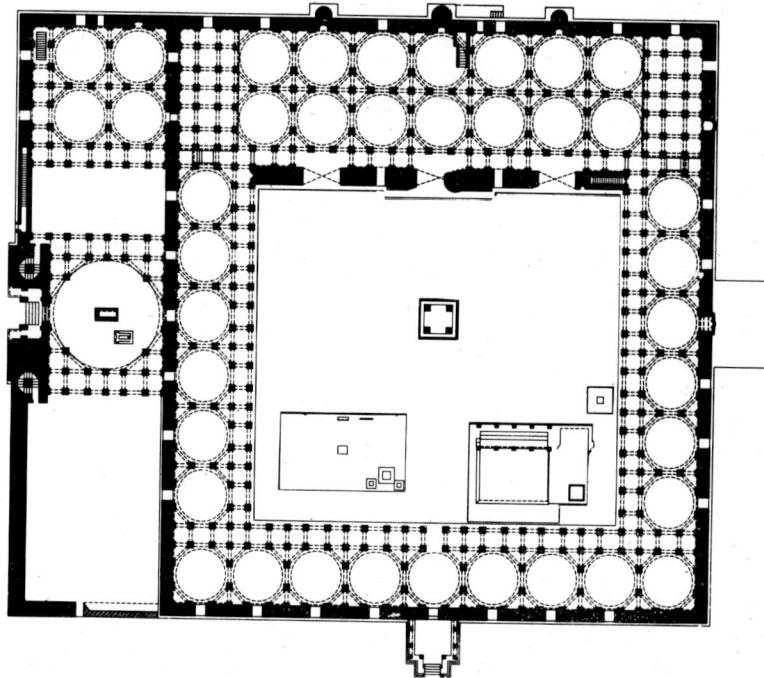

Mosques at Dholka

The scene of Muslim architecture in Gujarat now shifts to yet another place of importance and great wealth, popularly known today as Dholka (the ancient Dhavalakha of the twelfth century). It was earlier ruled by the same potentates of Gujarat—the Vastupula and Tejpala brothers. Here, the Muslim governor, Hilal Khan Qazi, in true Islamic tradition, decided to build a public mosque. The design of the *liwan is* based on the example at Cambay, with the convenient omission of surrounding cloisters, but the addition of a prominent entrance gateway *(Figs 4.04, 4.05)*. The design of this gateway is taken almost *in toto* from the *antralaya* or entrance *mandapa* of a Hindu temple, complete with *asana* or inclined seats along its periphery. Though it is crowned with a hemispherical dome instead of a typical temple roof, passing through this porch one is surprised to find oneself in the courtyard of a mosque rather than in the interior of a Hindu temple.

At Dholka was then constructed a mosque popularly known as the Tank or Taka Masjid, 'so called from a water tank which is close to the east entrance.' The plan in configuration is similar to the Adina Masjid at Pandua, the sanctuary side of

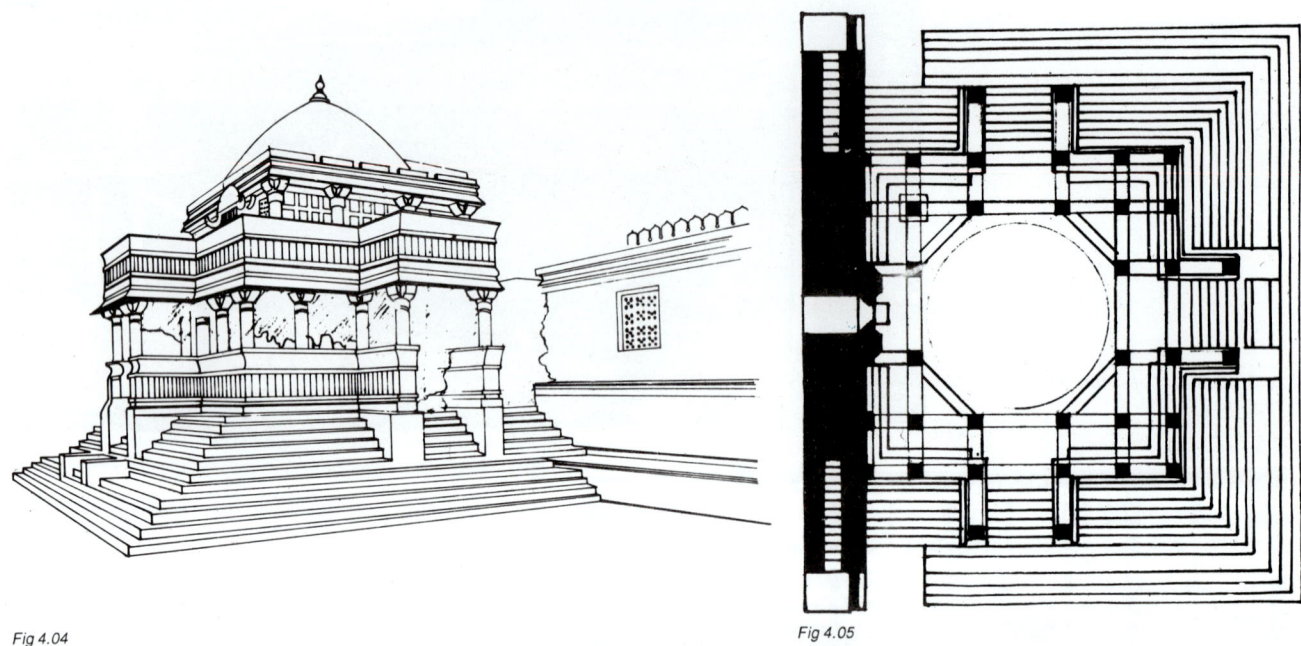

Fig 4.04   Fig 4.05

the courtyard being longer than the other sides. The mosque had little to contribute to the evolution of the Gujarati style, composed as it was of columns and beams reft from the Hindu and Jaina temples, of which there seems to have been an endless supply available.

Figs 4.04, 4.05 Sketch and plan of the entrance gateway of the mosque at Dholka, AD 1333

There was a long pause of over thirty years before any new production was forthcoming from the Gujarati builder. Having experimented with structures built either exclusively with material from Hindu temples or in juxtaposition with 'screen walls' of Islamic arches, the builders of Gujarat were as if waiting for a verdict on their initial efforts before proceeding further. This was soon to be given by the famous Gujarat ruler Ahmed Shah who declared independence from the Delhi Sultanate in AD 1391. Ahmed Shah's decree on architectural activity in his kingdom, preoccupied as he was with politics, seems to have permitted the Hindu builder to practise his traditional architectural and structural technique, devoid of figurative sculpture, but adequately embellished with Islamic motifs. The most prominent of the latter was the arched masonry wall in front of the *liwan*, adorned with *minars* or turrets.

Fig 4.06 A typical minar in the Gujarat style

Minars of Gujarat

In the early years of Ahmed Shah's rule when he established his new capital city of Ahmedabad in AD 1401 on the site of old Karnawati on the left bank of the river Sabarmati, building activity was limited to the revival of the rather dull style of the Cambay mosque. Thus, in the Ahmed Shah and Haibat Khan mosques of Ahmedabad, the only progress made seems to have been the addition of what may at best be described as *minar* turrets on either side of the central arch of the screen wall *(Fig 4.06)*. This was surely inspired by the contemporary Tughlaq mosques of Delhi where circular buttresses flanked the central arched pylon. And it must be said that the decorative design detail of the *minars*, an architectural form that he was not at all familiar with, utterly confused the Gujarat builder. The *minars* were 'thus pummelled by him into a succession of horizontally delineated mock balconies, bracket and pillar reliefs' capped by an incongruous miniaturized temple *mandapa* roof *(Fig 4.07)*. Unlike the Tughlaq, and more particularly the Jaunpur builders, he never quite comprehended the essentially vertical architectonics of this Islamic motif. The constructional techniques employed to erect it, too, were faulty, and only

Fig 4.07 Facade of Haibat Khan's mosque, Ahmedabad, early 15th century

a few complete examples of the Gujarat *minars* have survived in their complete form. The upper part of the *minars*, even of the modest Ahmed Shah's mosque, collapsed several years ago.

Building in Ahmedabad

The mosque of Sayyid Alam built in Ahmedabad in AD 1412 is the true prototype of the almost entire subsequent range of mosque architecture in Gujarat. The facade of this mosque 'contains several elements, such as porticos on the wings, projecting cornices and ornamental brackets, together with a variety of decorative details. 'In the interior, 'the appearance of the nave has been improved by an additional intermediate storey or embryo triforium.' In brief, with the completion of this structure, the Gujarati builder was at the threshold of undertaking more ambitious building projects. He received his first comprehensive commission from Ahmed Shah, who decided not merely to adorn his new city with a Jami Masjid but to infuse it with' an architectural environment appropriate to the imperial ceremonials.'

Such an environment was achieved by laying out a processional 'king's-way' that connected the gates of his palace to the northern side of the grand public mosque. The imperial avenue issued from a stately triumphal arch, the so-called Teen Darwaza *(Fig 4.08)* that was planted as an entrance gateway to the Royal Square of the citadel. Ahmed Shah was concerned with providing himself and his royal consorts with a grand, final resting place. For this purpose a virtual royal necropolis consisting of his own mausoleum and courtyard of the 'Tombs of the Queen' (or Rani ka Hujra) was laid out in front of the mosque. It is, however, the Jama Masjid that is the vibrant heart of this great town planning scheme.

Jami Masjid, Ahmedabad

The Jami Masjid of Ahmedabad, completed in AD 1423 and measuring 382 ft × 258 ft (116 m × 79 m) is generally considered the nadir of mosque design in western India. Judged cooly by abstract standards of Islamic architecture, this mosque, like almost all other works of Gujarati Muslim builder, may fare none too well. From a more subjective and almost romantic point of view, the building speaks of the silent fusion in stone of the souls of two religions that were apparently as contradictory as day and night: The fusion here, though, is not of one entity dissolving into the other, but rather, the two meeting on equal terms, lending and borrowing in a truly democratic spirit to make for a rich blend. What is indeed surprising, is that this architectural 'secularism' should manifest itself in Gujarat. It is on record that Ahmed Shah, the moving spirit behind the erection of this mosque, was one of the most bigoted Muslim rulers who indulged in wanton destruction of Hindu places of worship. Mercifully, though, Ahmed Shah failed to perceive that the building of the Jami Masjid in essence followed the 'scheme of constructing a temple building and introducing it into the mosque sanctuary as a central compartment.' If he had seen through this architectural stratagem of his builders, he may well have been impelled to order the destruction of his own *masjid*.

Fig 4.08 A view of the Teen Darwaza, Ahmedabad, AD 1425

In fact, the architectural theme that enlivens the entire design scheme of the Jami Masjid of Ahmedabad, is of contrast. This is apparent in its every part, particularly in its facade which, as a composition of solids and voids, is 'superb' *(Figs 4.09, 4.10)*. The solid, walled and buttressed central triple-arched composition is flanked by the airy lightness of peristylar verandas. Then again, even through the graceful Islamic arches of the central postern can be seen 'the alternation and interplay of light and shade among its frontal columns, and finally that engrailed arch springing so lightly and fancifully from its slender shafts.' If any doubt still persists as to the worthiness of this mixed architectural idiom one need only to enter the inner sanctuary. To the orthodox Islamic eye the interior may be 'garrulous and composed of lithic struts propped one above another like rivetted scaffolding,' but to another, the interior is 'as if the poetry of the temple had flown into the sobriety of the mosque' to create a mysterious spiritual equilibrium.

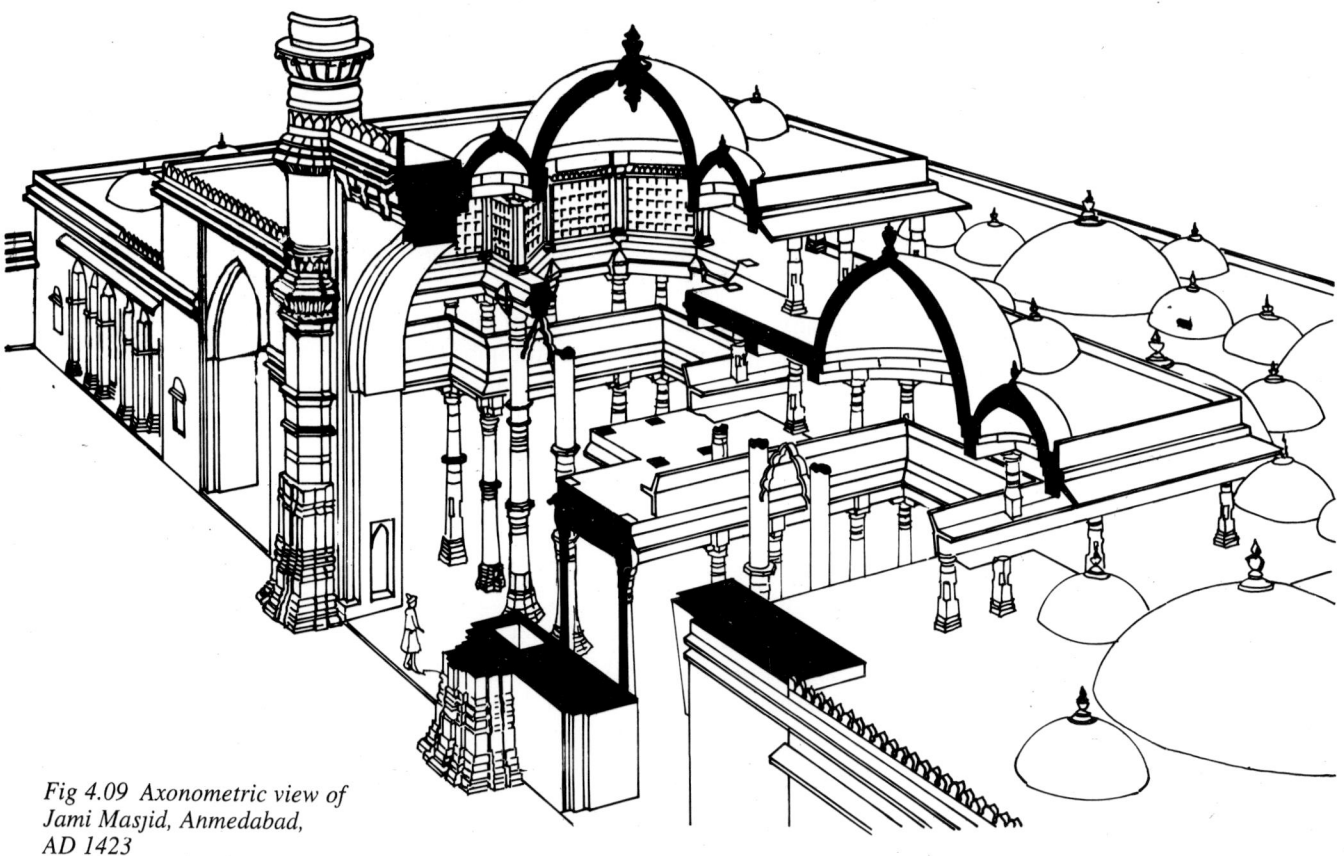

Fig 4.09 Axonometric view of Jami Masjid, Anmedabad, AD 1423

Liwan of the Jami Masjid

The 210 ft × 950 ft (64 m × 290 m) *liwan* of this Jami Masjid, instead of being a huge plain hall, consists of some 300 tall slender pillars, 'so closely set in parts that the interior simulates a thick grove of silverpine trunks'. The hall, however, does not evoke the deathly monotony of the South Indian halls of thousands of pillars, Islamic tenets, both in terms of space and construction, encouraged the designers to build in a central shaft of space rising vertically through two tiers of flanking balconies. This dynamic central volume is roofed by a large corbelled dome resting over an octagonal ring of columns, each of the eight facades being filled in with panels of *jaalis*. In one deft stroke of design, the builders had solved many functional problems of mosque design in Gujarat. The balconies provided a sufficiently private

Fig 4.10 The 'solids and voids' composition of the facade of the Jami Masjid, Ahmedabad

zenana apartment for the ladies; the domed roof added an Islamic quality to the interior as well as the facade; and the open grills the dome made the central bay an open shaft generating cool currents in the air, so essential in the hot and humid climate of Ahmedabad. The upper storeys of the two mysteriously shaking minarets, which may have been the only visually discordant notes of the facade, were destroyed in an earthquake.

While Ahmed Shah's *masjid* marks in many ways the culmination of Gujarati mosque design, his tomb furnished a model for the innumerable Gujarati tombs that followed. The plan, once again, is devised from the Hindu temple *mandapa*, reorganized to allow for a square enclosed and domed chamber in the middle to house the grave. Thus, the characteristic Gujarati tomb, instead of being composed of a solid cube and hemisphere, became a dome resting over a square open peristylar colonnade roofed with smaller cupolas.

The Garden Suburbs of Sarkhej

On the death of Ahmed Shah, his son Mohammed Shah was prompted by pious intentions to build a tomb in homage to a famous recluse, Sheikh Ahmed Khattri, who died in AD 1441 at Sarkhej, some six miles south-west of the city of Ahmedabad. The Sheikh had certainly chosen a site of considerable picturesque appeal, the focal point of which was a large water tank. In due course Sarkhej became a suburban retreat for the Gujarat rulers, and gradually 'there arose palaces, gardens, pavilions, gateways, and a large artificial lake, besides other mausoleums' *(Fig 4.12)*. The royal

Fig 4.11 View of Hindu columns bunched together at Sarkhej

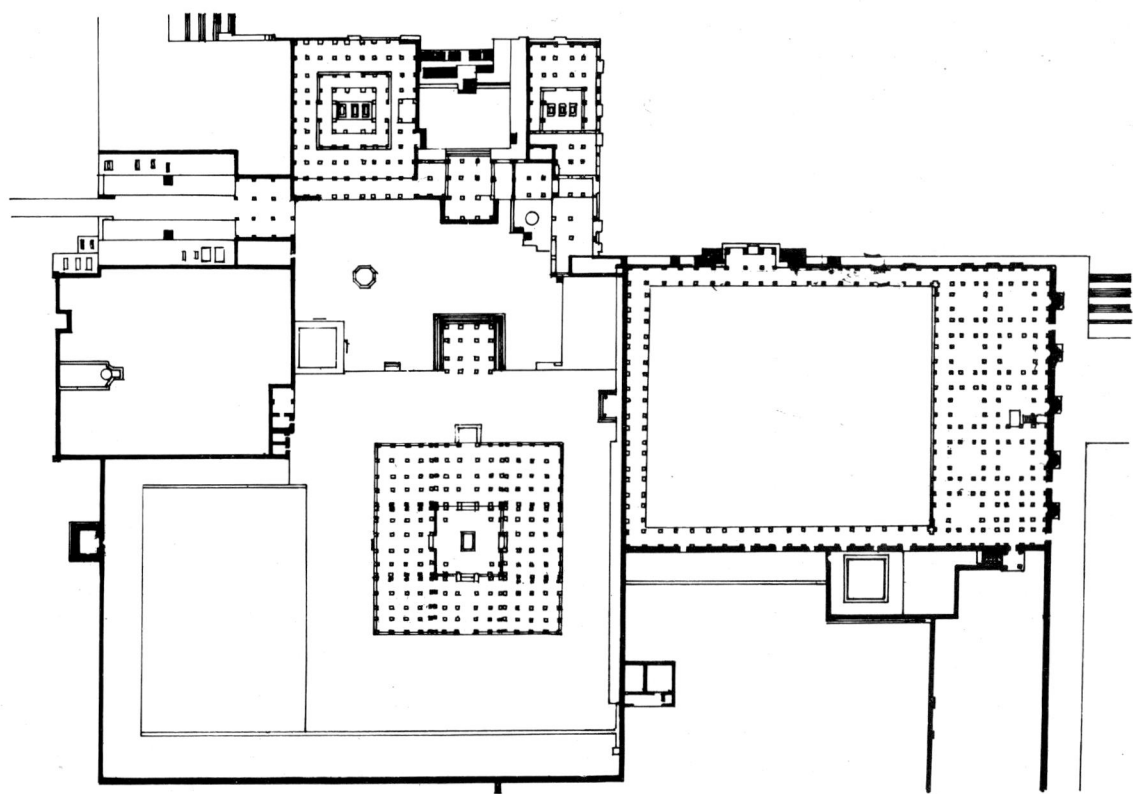

Fig 4.12 Layout plan of the garden suburbs of Sarkhej near Ahmedabad, AD 1451

usurper of Sarkhej began, however, with the building of a tomb and a mosque in honour of the Sheikh. The appeal of both these structures lies in their sumptuous largeness rather than fine architectural detail. In fact, the vast mosque, measuring some 225 ft × 157 ft (68 m × 48 m) has a rather wild appeal in the bunching together of Hindu columns of unending variety *(Fig 4.11)*. The equally huge tomb consisting of a square hall of 104 ft (32 m) side is an unusually large hypostyle hall forested with columns. The central domed portion of the hall has been enclosed within an unusual brass framed screen to form the sepulcher. Though later Mahmud Beghara decided to build his own tomb amongst these picturesque surroundings, it is the small pavilion in front of his tomb that best represents the architectural style of Sarkhej. This open pavilion is structured out of 16 tall and graceful shafts, devoid of any ornamentation except a countersinking on the angles. It is roofed with nine small domes, which both internally and externally form 'as pleasing a mode of roofing as ever was applied in such a small detached building of this class' *(Fig 4.13)*.

Fig 4.13 Pavilion in the Sarkhej complex

Brick Architecture in Gujarat

Soon after Mohammed Shah's death, during Qutb-ud-din' reign were built the Alif Khan mosque at Dholka and the Darya Khan tomb in Ahmedabad. Both are massive structures entirely fabricated out of brick. The use of brick as a building material necessitated that the structures be constructed in the Islamic arcuate style, entirely foreign to the delicate Hindu-Islamic balance of the Gujarati style. It is conjectured, too, that these structures built by courtiers of the rulers, were designed by architects from southern Persia who had been attracted to Gujarat by its fabulous riches. Since these two buildings did not make a marked impression on the mainstream of Gujarat Islamic architecture they are best left alone for the records of the archaeologist, than described in detail as a part of the Gujarat tradition. Suffice it is to say that while the Dholka mosque *liwan*, consisting of three conjoined square chambers flanked by square towers on either side, shows an influence of the Tughlaqian buttresses, the single chamber of Darya Khan's tomb surrounded by a domed passage all around, is but a robust version in brick of the traditional Gujarati tomb. Further, conjectural details of these rather dilapidated structures, particularly Alif Khan's mosque, would only add to the confusion in understanding the innumerable examples of the Gujarat craftsman's art. The confusion would be even worse confounded if one were to attempt to describe each and every subsequent structure of the Gujarati style. Once Gujarat, by various political strategems, came under the sway of the famous Mahmud Beghara, there was almost a frenzy not only of building but also town planning activity. Many new cities were laid out during the reign of Beghara, the most prominent of which was his new capital city of Champaner.

The 'Well Retreats' of Ahmedabad

In Ahmedabad, on the other hand, the builders seem to have been building not only mosques and tombs in the city and its suburbs of Batwa and Usmanpur, but also numerous *baolis* or 'well retreats' as shelters from the heat of the long summer months. These hot weather retreats were not just civil adjuncts to the religious architecture, but veritable works of art in themselves. In fact, these so-called *wavs* of Gujarat 'were not merely erections over the well shaft but took the form of extensive subterranean galleries of a highly architectural order.' They generally consisted of a well shaft from which water was drawn up by ropes in the usual manner, and also a flight of steps which took one down to the water level. Around the shaft were arranged a series of underground compartments one over the other, 'the pillars, capitals, railings, wall surfaces and cornices all being as profusely sculptured as the temples and mosques of Gujarat.' Two of the more prominent examples of such well retreats are the Bai Hari at Ahmedabad (AD 1499) and the Bhamaria well just outside the town of Mahmudabad. Bai Hari's wav *(Figs 4.14a, b)* is approached through a fine domed pavilion from where one descends down an 18 ft (5.5 m) wide flight of steps. After passing through four levels of pillared loggias, each exquisitely crafted, one reaches

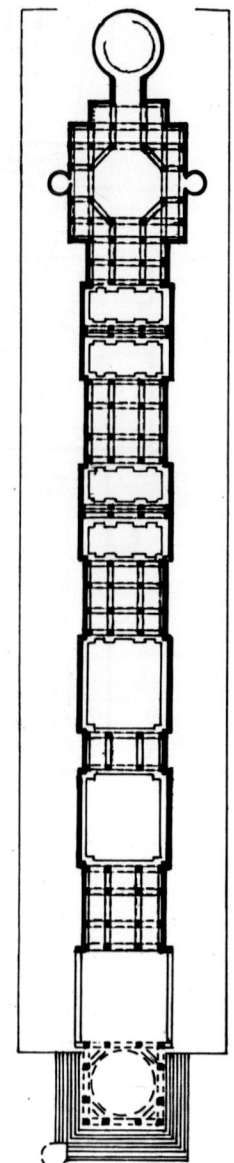

Fig 4.14a

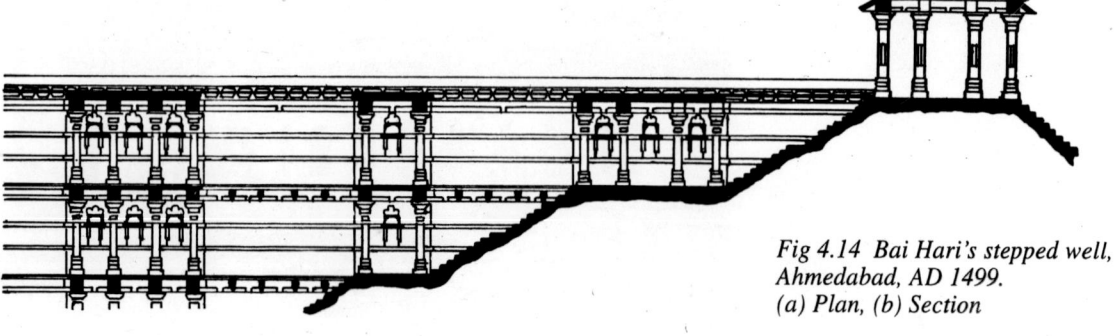

Fig 4.14b

Fig 4.14 Bai Hari's stepped well, Ahmedabad, AD 1499.
(a) Plan, (b) Section

Fig 4.15 View of Bai Hari's stepped well

the 24 ft (7.3 m) square central shaft from where one is led to the variable water level by two spiral staircases. Four tiers of pillared galleries support the sides of the shaft and provide cool resting places for the people using the well *(Fig 4.15)*. The Gujarat craftsman, to whom the building of a well was just as sacred as that of building a temple or mosque, had thus erected a fine treasure house of culture for those who came to draw water daily from the well.

The Bhamaria well as located in the centre of a pleasure garden. According to local tradition, 'the emperor had it constructed as a summer retreat, and the two stone arches over it were erected to hang the King's swing upon.' Apart from the circular chambers erected around the octagonal well at ground level, four flights of staircases led down the eight subterranean chambers. From here, four spiral staircases led down to chambers closer to the water level in the well. So long as the water in the well was kept fresh nothing would be cooler during the heat of the day in early summer than these rock-hewn chambers. The courtiers had their consorts obviously spent hours here looking down from the ornate balconies and arched windows into the cool and still waters of the well.

Fig 4.16a

Fig 4.16b

The Rauzas of Gujarat

Taking the cue from Ahmed Shah's idea of building his own mausoleum adjacent to the Jami Masjid, the ruling family of Gujarat rationalized the concept into that of a *rauza*. In its purest form, the *rauza* consisted of an arrangement in which the 'tomb and its mosque confront one another, and being complementary in design, together produce an attractive composition.' The strategy of combining the tomb and mosque into a unified whole also ensured that the glory of the ruler would be duly preserved. Depending on various circumstances, sometimes the mosque and sometimes the tomb would dominate the architectural appeal of the *rauza*. For example, in the *rauza* of Sayyid Usman at Usmanpur, a suburb of Ahmedabad on the further bank of the Sabarmati, it is surely the 100 ft (30.5 m) square mausoleum that is more worthy of note than the mosque. The most noteworthy feature of this tomb is the inner ring of twelve instead of the conventional eight columns to hold up the central dome. The central enclosure is surrounded by two concentric aisles formed by bunching columns in clusters of four. At each of the corners is planted a small dome to complete the *panchatayana* form of planning *(Figs 4.16a, b, c, d)*. Apart from an odd example here or there, this was also the last of the Gujarat tombs to use the Hindu trabeate system for its construction. In fact, it was obviously the rather confused layout of its columns that prompted the builders to rationalize their planning methods.

Fig 4.16 Rauza of Sayyid Usman, Ahmedabad. (a) & (b) Views, (c) Plan, (d) Elevation

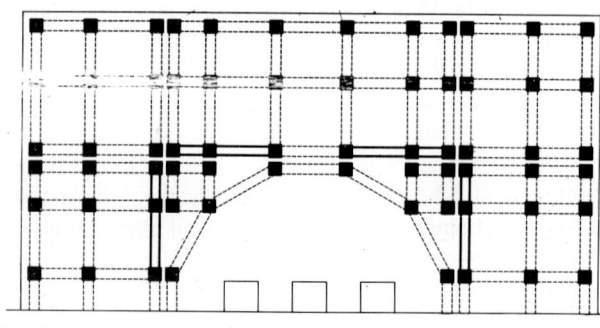

Fig 4.16c

Fig 4.16d

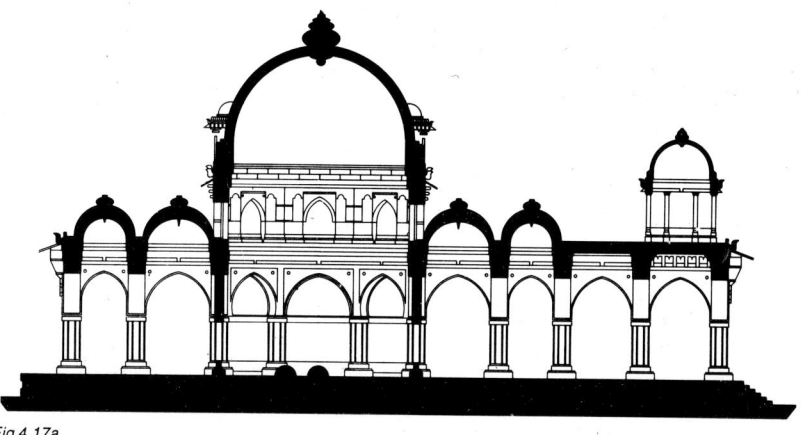

Fig 4.17a

Islamization of Gujarat Architecture

Fig 4.17 Tomb of Mubarak Sayyid near Ahmedabad, AD 1484, (a) Sectional elevation, (b) Plan

By the middle of the fifteenth century, like their Muslim colleagues in other regions of India, the Gujarat rulers, too, were impressing upon their architects the need to adopt the more Islamic arcuate style of building. Thus, in the mosque of Achut Kuki (AD 1472), though the interior continued to be assembled on the Hindu temple style, the trabeated side flanks of the Ahmedabad Jami Masjid on which this mosque was modelled, have been discarded. Obviously, as a part of this sustained policy of Islamization, in the small and rather ungainly mosque of Muhafiz Khan, the elevation is reduced to a flat three-arched facade. Minars have been planted at the two extremities, and the Hindu balconies relegated to the sides and rear. This preference for the arcuate style was more successfully expressed in the tombs of Qutub-u-Alam at Batwa (AD 1480) and of Mubarak Sayyid near Mahmudabad about 17 miles south-east of Ahmedabad. While the Batwa example is a Muslim arcuate skeleton built up over a rationalized Hindu temple *mandapa* plan, the tomb of Mubarak Sayyid is a more handsome and convincingly Islamic tomb. The 95 ft (29 m) square plan configuration of this tomb, consisting as it does of a group of definable clusters of four pillars, is specially designed to receive a dome over each of its square bays *(Figs 4.17a, b)*. The central 36 ft (11 m) diameter dome, rising

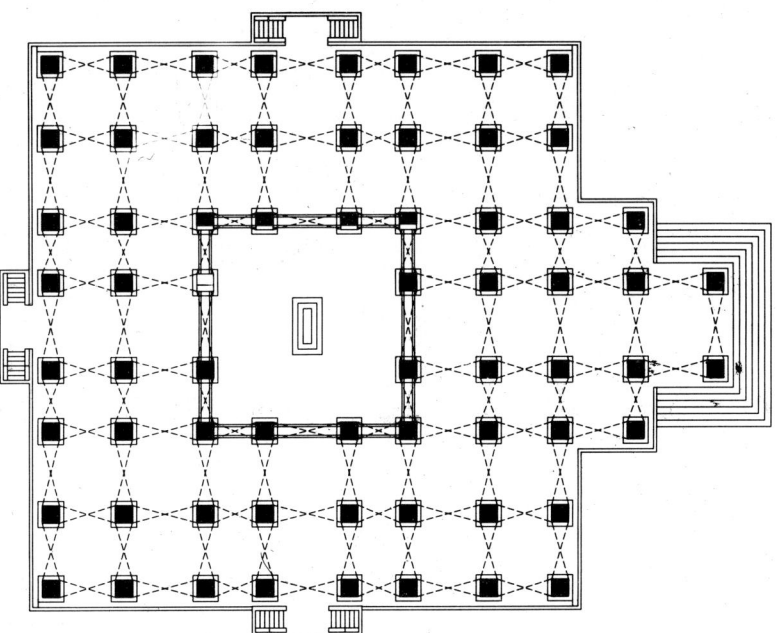

Fig 4.17b

to a height of almost 70 ft (21.3 m), rests over a central square defined by 12 columns, and is cordoned off as the inner sepulchre by *jaali* infill panels. The inner sanctum receives due definition on the exterior, too, by the planting of four *chhatris* at the cardinal points of the dome. In fact, it is apparent that the tomb is the product of Gujarati craftsmen guided by an architect from Delhi familiar with the more developed tombs of the Lodis. The only design concession that the Hindu craftsman squeezed out from his Muslim overseers was the planting of an entrance canopy attached like an *antralaya* to the otherwise perfectly square plan of the tomb.

Fig 4.18 Jami Masjid, Champaner, AD 1485

Beghara's Capital at Champaner

Meanwhile, Mahmud Beghara had captured the fort of Champaner, situated some 71 miles south-east of Ahmedabad, from the Hindu chieftain Jaysingh Patai Rawal. He was obviously fascinated by the picturesque 2,500 ft (762 m) high fortified hill of Pavagadh surrounded by forests of the plains below, and decided to build a new capital city in Champaner. Mahmud Beghara's own great city ultimately added more to the architectural wealth of Gujarat. The city, which took a quarter of a century to build, was only inhabited during his time, and today, 'passing along its silent grassgrown streets from one noble monument to another,' the remains convey 'the impression of a shimmering mirage which, on close acquaintance will dissolve into nothingness.' However, of the numerous remnants, some of which are now surrounded by the forest, two major building sites have not so far 'dissolved into nothingness.' These two, the Jami Masjid and the Nagina Masjid, adequately represent the architectural style of Champaner.

Champaner's Jami Masjid

The Jami Masjid is the more significant of the sites of the city and was undoubtedly inspired by its counterpart at Ahmedabad. The marked design development in the style of the Champaner *masjid* over its counterpart at Ahmedabad has two distinct aspects. First, due care has been lavished on the ancillaries attached to a *liwan* to make it a more comprehensive mosque plan *(Fig 4.18)*. Second, the decorative aspect of these ancillaries tends, if anything, to be even more Hindu than those at Ahmedabad. Thus, though the facade of the *liwan* discards the open colonnade wings of the Ahmedabad prototype, the other three sides are ornamented with balconies, bracketed openings, turrets, buttresses and corner *minars* singularly Hindu in ornamentation. Moreover, in the interior, the concept of a 'temple inside a mosque' is as predominant as in the Jami Masjid at Ahmedabad. It is composed of a hypostyle hall of a forest of columns piled up one over the other *(Figs 4.19, 4.20)*. In the central

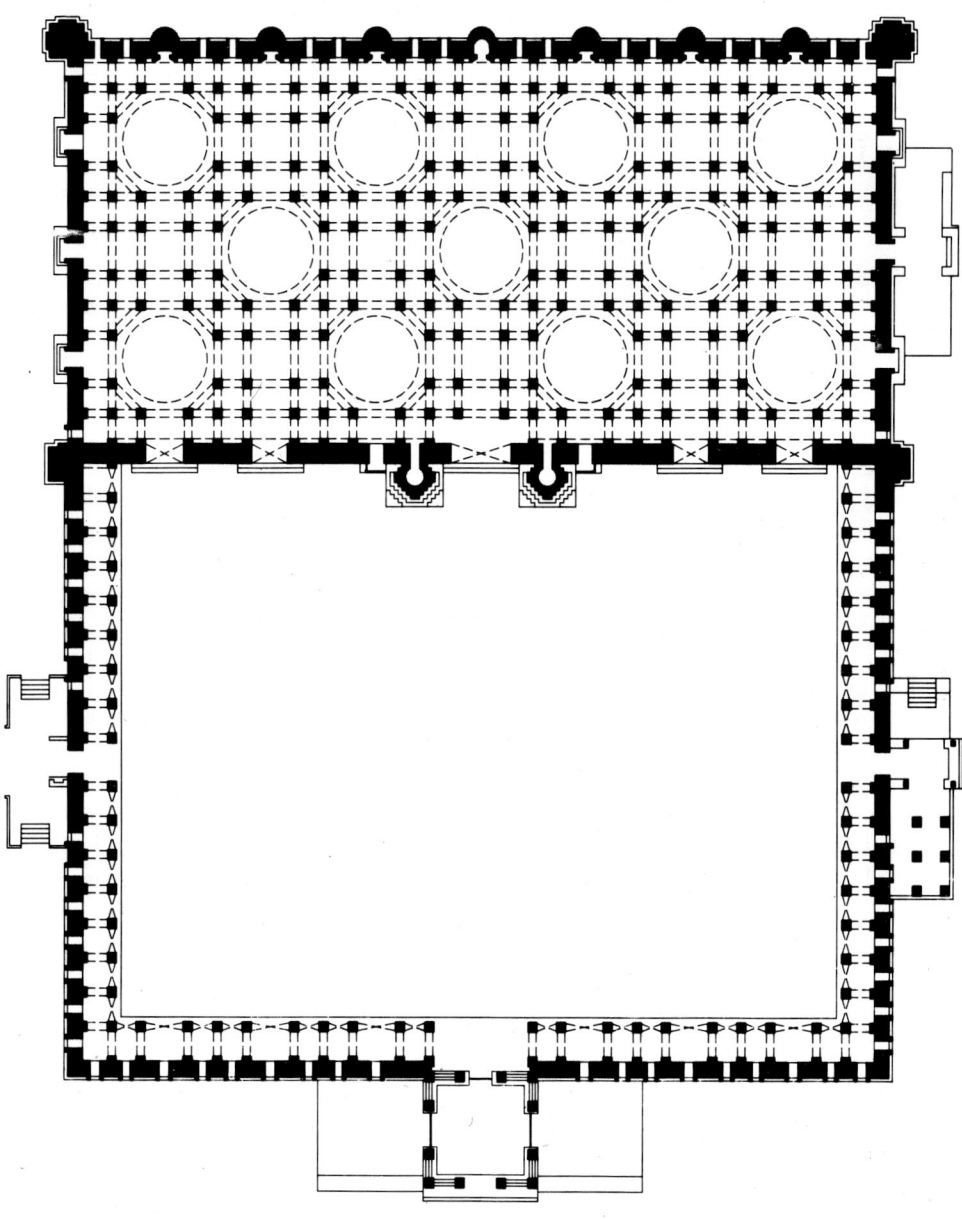

Fig 4.19 Plan of Jami Masjid, Champaner

Fig 4.20 Cut-away perspective of Jami Masjid, Champaner

part, it achieves the effect of a domed nave rising to a height of 65 ft (20 m). On the top storey, the balcony enclosing the rotunda is octagonal with a ribbed and richly fretted dome rising on pillars. The galleries themselves, separated from the pillared prayer hall below, provide retreats for peaceful meditation. The entire layout does not achieve the varying heights in the ceiling but is more rational than that of the Ahmedabad *masjid*.

It is only the entrance gateway to the Champaner mosque that gives a hint of the fact that the 'tilt in policy' towards an arcuate Islamic style in Ahmedabad had not entirely escaped the notice of the builders of Champaner. In fact, it would seem that the gateways and colonnades that enclose the *sahn* were added well after the construction of the *liwan* had been completed, by which time more orthodox Islamic ideas from Ahmedabad had reached Champaner. And so, the entrance gateway is designed as a square chamber with each of its walks composed of three symmetrically arranged, firmly contoured arches filled in with *jaali* panels. The cubic base is sheltered by a deep *chajja* projecting over closely spaced brackets. Above this markedly Hindu element is a broad parapet, at each corner of which is planted a domed *chhatri*. Built up over the square chamber the dome was supported by the decidedly Islamic method of corner squinches, but has now vanished.

The entire decorative scheme is purely geometric, with just a dash here and there of the flowing forms of Hindu sculptural details. That the design of the gateway of the Jami Masjid became the prototype of the Champaner architectural style is apparent from the remains of the so-called Nagina Masjid tomb, of which the dome and parapet have also collapsed. The sub-structure is only a slightly more florid and richly ornamented version of the Jami Masjid gateway.

The Jaalis of Siddi Sayyid at Ahmedabad

In spite of Beghara's valiant efforts to establish a new capital at Champaner, Ahmed Shah's city of Ahmedabad remained the pivot of the Gujarat Muslim empire. While the shimmering glory of Champaner vanished just as quickly as it had been created, the city on the banks of the Sabarmati continued to entice the Gujarati builder. Thus, in the Siddi Sayyid Mosque, built here in AD 1515, the builders for the first time suggested that the entire facade of the mosque *liwan* could be installed. The customary minarets in such a frontage were necessarily relegated to the northern and southern extremities. The structure, though a sober and handsome example, is by no means great architecture. But almost as if out of the blue, there appeared a master craftsman on the scene who converted the *jaali* infill panels in the arched aperture in the western walls of the *liwan* into masterpieces of art *(Figs 4.21, 4.22)*. It was as if some genius responded simultaneously to both the Islamic requirements of non-

Fig 4.21 Converted jaali infill panels in the arched apertures of Sayyid mosque, Ahmedbad, AD 1515

Fig 4.22 Stone jaali in the rear wall of the Siddi Sayyid mosque, Ahmedabad

figurative geometric decoration, and the Hindu ingenuity for producing amorphous and ambiguous illusions from an extremely rational and logical framework. The screen in the mosque, though eschewing all figurative elements, is animated by a sensuality that belongs only to the female human form. For expressing this sensuousness the artist has chosen the entwining trunks, branches and leaves of trees and plants as his subject. The sinews of this natural element are then minutely dispersed within the arched outline in a manner suggestive of elementary simplicity at first glance, and extreme complexity on deeper study. In a sense, this one piece of ornamentation is as attractive to the rustic viewer who can but marvel as its intricate craftsmanship and sensuousness, as to the seasoned designer to whom its almost casual reconciliation of 'Hindu-Muslim,' 'formal-informal' and 'crystalline-amorphous' ideals make it worthy of deep study.

Mosques of Rani Rupvati and Shah Alam

Finally, the builders in their glorious isolation from Islamic architectural ideas in other parts of India and, as if to sum up the philosophy of Gujarati architecture in one last great statement, produced two more significant monuments. These are the mosque of Shah Alam (AD 1550) and the mosque of Rani Rupvati (AD 1515). The

Fig 4.23 View of long arcuated facade of the mosque of Shah Alam, AD 1550

Fig 4.24 Rani Rupvati's mosque, AD 1515

former, in its austerely articulated, long arcuated facade and wide open courtyard, speaks of the supreme glory of the abstract brotherhood of Islam *(Fig 4.23)*. The latter, as eloquently, makes out a case for the sublime reconciliation of the apparently diametrically opposed ideals of Islam and Hindu architecture *(Fig 4.24)*. And the wonder of Gujarati's Islamic architecture lies in the fact that right up to its dying moments the two great ideals were equally alive and invigorating. The facade of Shah Alam's mosque (if one were to ignore the incongruous minarets), is indeed a refreshing, honest and almost hygienic representation of Islamic ideas. Rani Rupvati's mosque, on the other hand, liberally incorporates deep *chajjas*, pummelled *minars* and temple balconies in the arrangement of the central facade in a manner evocative of the Hindu's undying genius for assimilation, and for artistically carving folk motifs. It is only in the Muslim architecture of Gujarat that the physical 'summing up' is so truly symbolic of the Hindu and Islamic pulsations that had throbbed life into the monuments of Gujarat.

Mehtar Mahal,
Bijapur

CHAPTER 5

Bahmanis and Adil Shahis in the South

AD 1344–AD 1672

While builders in the northern, eastern and western provinces of India were evolving an Indian Islamic style by grafting one local tradition over another, the seeds were being sown in the South of a Muslim building tradition that flowered in striking and glorious isolation from the influences of its environment. In spite of the fact that Mohammed Tughlaq, during his mad venture of shifting the capital to south central India in AD 1338, had planted the earliest Muslim mosques in the South—the Jami Masjid at Daulatabad and the Deval Mosque at Bodhan—the architectural style of the Tughlaqs of Delhi did not take firm root in the southern peninsula. Of the earliest mosques, while the 206 ft (63 m) square Jami Masjid at Daulatabad was produced entirely with materials from temples in its vicinity, the one at Bodhan was merely a 'Jaina temple transformed by a few structural additions to do service as a mosque.' Also, it would seem that with Tughlaq's retreat to Delhi and the desertion of Daulatabad, the style, if it may be called one, of the buildings of this city did not play a prominent role in the development of Muslim architecture in the South. What is more, as we shall see, a good measure of inspiration for new ideas from Islamic countries across the Arabian Sea.

Architectural Inspiration from Abroad

Various circumstances led to the southern Muslim builder accepting novel building ideas from overseas Islamic countries rather than developing on indigenous sources. The insularity of the South Indian Brahmin and his master craftsmen did not encourage the Muslim rulers to depend on the skills of Hindu architects to build mosques and tombs for them. Also, for historical and political reasons, the Muslim cities of the South like Bidar, Bijapur and Golconda were not built around live and thriving centres of Hindu culture. As such, spoils in the from of readymade Hindu building materials for the fashioning of new structures were not abundantly available. Furthermore, mindful probably of the Hindu oriented Islamic architecture of Gujarat, the foreign blooded Muslim rulers of the South were not so keen to perpetuate another Gujarat style 'architecture' in their dominations. It would also seem that in the fourteenth and fifteenth centuries, voyages to and from India across the Arabian Sea had become fairly frequent, and a number of skilled immigrants not only from neighbouring Persia but also from distant Turkey, were only too happy to serve the burgeoning Muslim overlords of the South. Thus, ironically enough, it was the conservative southern Hindu stronghold that was injected with doses of an alien and virile architectural style. The energy generated by this infusion culminated in the building of the great Gol Gumbaz at Bijapur. This was as triumphant and bold a structural statement, as subtle and glorious had been exotic sculptural achievements of the Hindu builders of nearby Halebid and Belur. But before the builders of Bijapur could aspire to build the largest dome of the sixteenth century world of architecture, the fourteenth century builders of Gulbarga and Bidar were busy laying the foundations of a building style that was to literally rise to dizzy heights under the Adil Shahis of Bijapur.

Bahman Shah's Fortress of Gulbarga

These foundations were earliest laid under the patronage of one Ala-ud-din Bahman who in AD 1347, having thrown off his allegiance to Delhi, established his capital city at Gulbarga in modern Karnataka State. Bahman Shah, an adventurer from the Persian court, had thrown off the political as well as architectural yoke of Delhi. He surrounded his city of Gulbarga with 50 feet (15.2 m) thick fortress walls and a 90 ft (27.4 m) wide moat scraped out of living rock *(Fig 5.01)*. At regular intervals the outer defences were punctuated by bastions, barbettes and battlements of cyclopean order. It then appeared that Bahman Shah invited an architect from the north Persian town of Qazvin to offset the ponderous military architecture of his fortress of Gulbarga with models of a more refined architectural style. However, of 'all the pavilions, palaces and kingly halls' that the ramparts of Gulbarga may have contained at one time, only the Jami Masjid of the city stands today, 'isolated in the midst of a scene of devastating emptiness' *(Fig 5.02)*.

Jami Masjid, Gulbarga

In the design of this Jami Masjid, the master builder Rafi was not inspired by the traditions of his Persian home further north, but more apparently by the form of Muslim religious edifices in eastern Europe, and at the back of his mind may have been a 'domed and vaulted hall of the Basilica type.' In this extremely original and daring innovation for the design of the Gulbarga mosque the decided to abandon the open-to-sky central courtyard and replace it with an entirely roofed-in, domed and

Fig 5.01 View of fortress wall surrounding the city of Gulbarga, AD 1347

Fig 5.02 Jami Masjid, Gulbarga, AD 1367

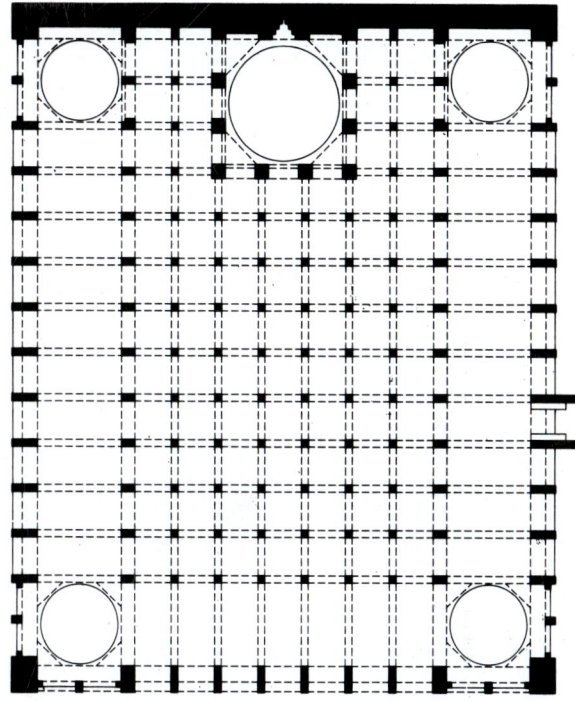

Fig 5.03 Plan of the Jami Masjid, Gulbarga

Fig 5.04 View of the mosque completely roofed with domes and barrel vaults

pillared hall *(Fig 5.03)*. Measuring a handsome 216 ft × 176 ft (65.8 m 53.6 m) and consequently covering an area of 37,916 sq ft (3,523 sq m) in plan, the mosque's central hall which is roofed over with 63 small cupolas, allows for 5,000 worshippers to get together for their Friday prayers. Four larger domes mark the corners of the periphery, while the sanctuary is crowned by a stately contoured stilted dome rising over a square clerestorey. The wide arcades on three sides of the central hall are spanned across by low arches, roofed over by long pointed barrel vaults that appear to form a cordon around 'the cratory of domed bays of the central hall whose thrust they sustained' *(Fig 5.04)*. The glorious white monument testifies as much to the

Fig 5.05 *A view of the square bays of the prayer hall of the mosque at Gulbarga*

originality of its designer's concepts as to his failure in establishing their acceptability. Though the Tughlaqian builders of Delhi did attempt a slightly watered down version of building mosques with semi-covered instead of open-to-sky courtyards, nevertheless a mosque with a covered courtyard did not find acceptance with the faithful for reasons elaborated earlier in the description of the Khirki Masjid at Delhi. The Qazvin designer, however, did not fail altogether. Many other architectural flourishes that adorn this mosque became part and parcel of the subsequent south Indian Islamic idiom.

Architectonics of Austere Dignity

Most prominent here was the raising of the stilted dome over a square clerestorey bay that in this instance served the functional purpose of allowing light into the otherwise enclosed sanctuary below. Again, the cloisters which are formed of a range of single archways of an extremely wide span and with unusually low imposts were repeated in many monuments of Bijapur. Lastly, and in many ways the most refreshing aspect of the entire edifice of the Jami Masjid is the boldness of its expansive and plain external surfaces. Its stilted dome, poised above the square substructure, 'though an affair of mass, has a light and aerial effect which is the result of strong yet refined contours and excellent proportion' *(Figs 5.05, 5.06)*. The harmonious and austere dignity of the architectonics of plain surfaces and meticulously organised geometric volumes must have been the prime reason why the south Indian Islamic builder eschewed the pulsating but sometimes over-ornamented surfaces of Hindu temples as inspiration for his mosques and tombs. Apart from stray cases of Tughlaqian influence from Delhi, this production of Gulbarga, as we shall see, had the most powerful influence in the development of the Deccani style of Islamic architecture.

Fig 5.06 *Arched cloisters of the Jami Masjid at Gulbarga*

Tombs of the Bahmanis

The pattern of tomb building had been established before architect Rafi built the Jami Masjid. And for the building of tombs the builders had no other models to follow except those of distant Delhi. Thus, the earliest group of tombs (including that of the founder of the dynasty who died in AD 1358) consisting of 'battered walls, sunken archways, heavy battlemented parapet, fluted corner finial and a low dome' is clearly derived from the Tughlaq prototype of Delhi. It is in the later group of Bahmani tombs which is popularly known as the Haft Gumbaz or literally 'one week (seven) domes,' that marks of Deccan originality appear. The batter of the walls is now hardly noticeable, and the 40 ft (12 m) high surfaces are demarcated into two storeys by blind, screen-filled archways. The arch itself with its well defined impost, imbibes something from the Gulbarga mosque. Again, as in the Gulbarga mosque, a new but consequently unsuccessful planning idea was adopted in the layout of the tombs. Thus, the tomb of Taj-ud-din Firoz *(Fig 5.07)* like others before it, consisted of two square conjoined chambers to form a rectangle measuring 78 ft × 158 ft (23.7 m × 48 m). The idea behind this 'double mausoleum' was that while one chamber was reserved for the cenotaph of the ruler, the other was meant to accommodate those of his family. This innovation, too, remained a part of the local Gulbarga tradition. It was not adopted by the succeeding line of kings when Ahmed Shah transferred the capital to Bidar, over 60 miles (96 km) north-east of Gulbarga, in AD 1425.

Fig 5.07 The double dome of the tomb of Taj-ud-din Firoz, Gulbarga, AD 1422

The Fortress and Mahals of Bidar

Whatever other political reasons there may have been for a shift of capital, Ahmed Shah no doubt was also attracted by the 'picturesqueness of the 2,330 ft (708 m) high and healthy plateau' on which Bidar was located. From the nature of the extensive works carried out in the building of a two-and-a-half mile (4 km) long, 50 ft (15 m) thick wall of laterite and trap *(Fig 5.09)* surrounded by a 30 ft (9 m) deep and 115 ft (35 m) wide triple moat, it is clear that the new Bahmani rulers were taking no chances with their new capital in this time of constant warfare. Each of the seven gateways in the wall *(Fig 5.08)* including the Mandu Gate of the citadel, was protected by draw-bridges and cleverly contrived to hold off any besieging army.

Within these grim bastions, however, there seemed to have flowed at one time a life of sensual enjoyment and luxury. Ahmed Shah's lifestyle is spelt out in the inscriptions on his mausoleum: 'Should my heart ache my remedy is that—a cup of wine and then I sup of bliss.' Even from the derelict remains of the number of fanciful *rangin mahals, zenana mahals, gagan mahals*, water palaces and hammams, it is apparent that he surrounded himself with all the artefacts to 'sup of bliss' after his cup of wine in the harems of Bidar. Unfortunately, now after more than 500 years of woeful neglect, it is only a few religious structures that give a clue to the more sober aspects of the religious architecture of Bidar. The sobriety of the architecture of mosque is in vivid contrast to the vivacious fancifulness of the secular architecture. While the one represents the colourful pageantry of the court, the other reflects the simple solemnity of the creed.

Fig 5.08 Gateway of Bidar Mahal, AD 1347

Fig 5.09 Fortress wall, Bidar Mahal

Jami Masjid, Bidar

The so-called Sola Khumba or 'sixteen pillars' mosque at Bidar is indeed representative of this solemnity, but of little else. It is built in a sedate and unaffected style consisting of a 77 ft (23.46 m) wide and 295 ft (89.8 m) long sanctuary *(Fig 5.10)*. The *mehrab* in the middle of the *liwan* is enclosed within a square compartment over which rises a stilted Gulbarga-type dome resting on an octagonal base. The only curious aspect of this otherwise humdrum mosque is the installation of three scores of 4.25 ft (1.3 m) diameter, plain, circular pillars to hold up the groined roof of the long chamber. The only other instance where circular pillars have been used in Islamic architecture in India was in the so-called Hawa Mahal in the Malwa region. But the more exotic and fanciful of the structures at Bidar is the somewhat better preserved *madrassa* of Mohammed Gawan, Malik-u't-Tajur, Prime Minister, scholar and military general of Mohammed Shah III. This scholarly Persian, who hailed from Gilan on the Caspian Sea, spared no effort, including that of importing workmen and even the essential building materials, to achieve his purpose of raising in Bidar a perfect copy of the kind of building in which he must have received his early instructions in his native land. He succeeded eminently in this venture.

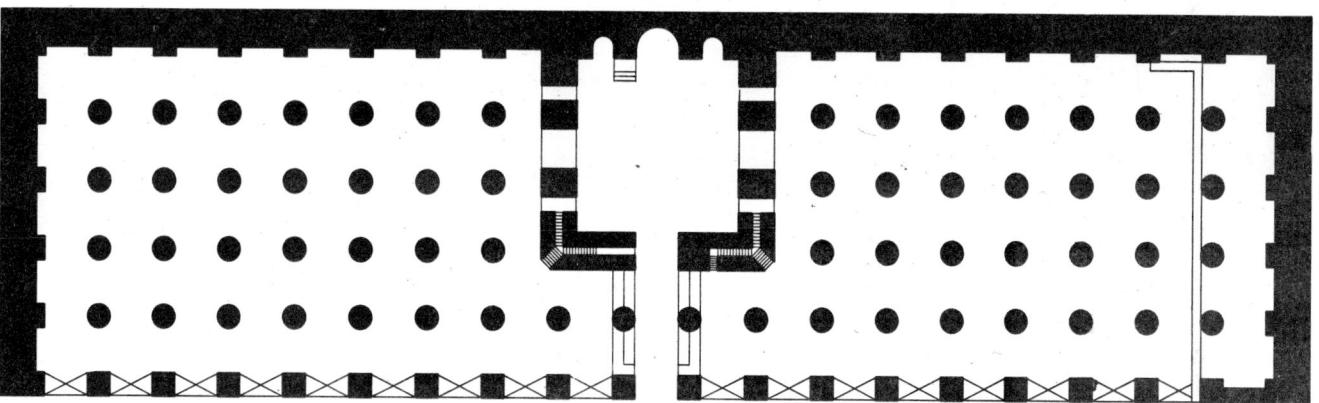

Fig 5.10 Plan of the Jami Masjid at Bidar

Madrassa of Gawan

Even in its present ruinous state, the building popularly known as the Madrassa of Gawan seems 'like a piece of Persia planted in India.' The building plan of a typical Persian university, without the slightest modifications whatever to suit the new environment was used in Bidar. Built sometime in AD 1481, this 205 ft ×180 ft (62 m × 55 m) rectangular three-tiered structure consists of a series of lecture halls, library, mosque and professors' and students' rooms arranged around an open-to-sky central courtyard measuring 100 ft (30.4 m) square *(Fig 5.11)*. The entrance facade on the east emphasized by two 100 ft (30.4 m) high Persian minarets on either side of a lofty gateway, while in the middle of each of the other three sides are planted semi-octagonal shaped bastions crowned by a typical Tartar dome. The rest of the outer surfaces of the *madrassa* are three rows of deep and severely formed arched niches, devoid of any other sculptural embellishments, recesses or projections. In fact, the entire structure was specially designed in the true Persian tradition as a series of vast flat surfaces, all to be covered by a layer of brilliantly glazed tiles. Every part of the facade was overlaid with patterns obtained by this method, primarily in green, white and yellow. To protect these exquisitely arranged floral, arabesque and even inscribed tile arrangements from damage by the damp of the earth, the builders laid layers of lead in the lower courses of the masonry. But alas, all these precautions proved to be a

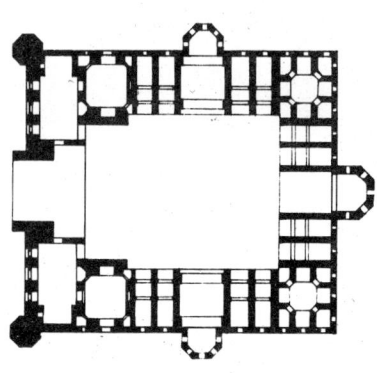

Fig 5.11 Plan of Gawan Madrassa, Bidar, AD 1481

Fig 5.12a

Fig 5.12c

failure. In the later use of the building as any army barrack, an explosion blew up a large part of the building including the entrance gateways and one minaret. Nature has done the rest and today, the remaining crumbling structure is but a pale shadow of its once brilliant exterior *(Figs 5.12a, b, c)*.

Barid Shah and the Lotus Dome

Under the Barid Shahi rulers who occupied Bidar for over 100 years from AD 1487 onwards, the city acquired a vast necropolis composed of tombs of the various Barid rulers. The architectural composition of all of these is simple in itself consisting as it does of an almost open cubic chamber below and a dome above. The cube below, in the now well-established popular technique, received long bands of coloured glazed tiles with inscriptions in the Kufi, Tughra or Nakshi script containing personal eulogies and sacred couplets. It is, however, in the construction of the dome that the Barid Shahi builders took a step in a direction that was to reach its climax in the building of the Taj Mahal. The profile of the later domes of Bidar took the form of a slight construction in the lower contour or an inward return of its curve near the base. This novel form for the dome was most likely suggested as much by the architectural experiments of Timur in Samarkand and the Safavid domes of Persia, as it was a resurrection by the Indian craftsmen of the ancient lotus arch and dome of the Buddhists *(Fig 5.13)*. It is certain, however, that this form of the dome suddenly aroused the enthusiasm of the Indian builder to embellish it with motifs definitely inspired by the classic architecture of the Hindus and Buddhists. In the subsequent domes of this nature the traditional petals at the *griva* or neck and the *maha-padma* at the crown of the dome made of the resemblance between the lotus dome of the Buddhist and the so-called onion, bulbuous or Tartar dome of Islamic Indian architecture, more than a matter of mere accident.

Fig 5.12b

Fig 5.12 (a, b) The brilliant exterior and (c) crumbled structure of Gawan Madrassa

The Tombs of Golconda

With the downfall of Gulbarga and Bidar as centres of power, the appointed governors in the nearby regions of Golconda and Bijapur declared their independence from the parent dynasties to establish the new dynasties of the Qutb Shahi and the Adil Shahi rulers respectively. Outside the fabulous and legendary city of Golconda near modern Hyderabad, is another necropolis *(Fig 5.14)* more vast than that of the Barid Shahi kings of Bidar. In its outer formations, the architectural style of the tombs is in direct succession to the tombs of Bidar. In the building of these tombs the Indian craftsman's penchant for sculptural embellishment and his traditional 'horrow vacuii' seemed to have found adequate vent. Apart from the ornamentation of the neck and crown of the dome with the petal motif and with the *maha-padma*, the surfaces of the lower cube, too, are richly ornamented in cut plaster. From the corners of the lower mass rise short minarets in the *guldasta* motif. The merlons along the parapet, too, are no longer stiff and militant but richly convoluted. The Indian builders here broke through the mass of the cube and hemisphere by providing a veranda running around the cubic half, thereby giving to some of the tombs the impression of a two-storeyed structure. The zenith of the Golconda style, however, was not the tombs but a ceremonial gateway popularly known as the Charminar of Hyderabad *(Fig 5.15)*. In intent and function much like the Teen Darwaza of Ahmedabad, it is a substantial 100 ft (30.4 m) square structure, the four corner minarets of which rise to a height of 186 ft (57 m). Each of the four facades contains the familiar ogee arch of 36 ft (11 m) span, above the apex of which are diminished storeys at arched and richly embellished cornices. Though the entire conception is inventive and spirited, there is also in it shades of ostentation, indicating the path of Hyderabadi architecture towards a sort of decadence.

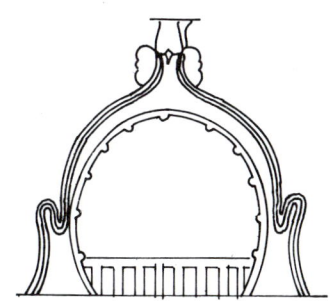

Fig 5.13 Buddhist arch—the source of the onion dome developed by the Barid Shahis

Fig 5.14 A tomb in the vast
necropolis of the Qutb Shahis at
Golconda

*Fig 5.15 The Charminar,
Hyderabad, AD 1591*

However, it was not the Qutb Shahis of Golconda, for all their artistic, intellectual
and cultural fervour, who really took the Islamic architecture of the Deccan to its
pinnacle of glory. This was to be achieved by the Adil Shahis who went about the
task of building their new capital city of Bijapur with zeal and enterprise unequalled
in the southern Deccan.

The City of Bijapur

Yusuf Adil Khan, a Turkish protege of the famous Mahmud Gawan of Bidar, and
Governor of Bijapur under the Bidar kings, asserted his independence from Bidar
soon after his mentor had been unjustly murdered by Mohammad III in AD 1481.
Inspired by the exemplary zeal of Gawan, Yusuf Adil set up a court at Bijapur that
was singularly free of any signs of bigotry, and many 'learned men and valiant officers
from Persia, Turkestan and Rum, and also several eminent artists lived happily under
the shadow of his bounty.' He himself was responsible only for building the walls of
the citadel, a fortress irregularly circular in plan, a few imperial buildings and two
small mosques prepared from despoiled temples. However, he seems to have inspired
in his successors 'a structural ardour that resulted in Bijapur being among the few
cities that have the most profuse display of fine buildings in India.'

Suburbs, Civic Amenities and Town Planning

It must be said right away, however, that apart from a cluster of fine building 'all of which were erected in a short burst of energy and speed within just a hundred years (from approximately AD 1550 to AD 1660), the rulers seemed to have had no inclination for any sort of systematic and appreciable concept of town planning *(Fig 5.17)*. Although the six gates in the six mile (9.6 km) circumference of the city wall are connected to the heart of the city by radial roads, there is no definite alignment or pattern in these arteries of thoroughfare. Rather, they seem to have expanded in a haphazard manner with the growth of the city, and the need was felt by successive

Fig 5.16 View of Chand Bauri, Bijapur

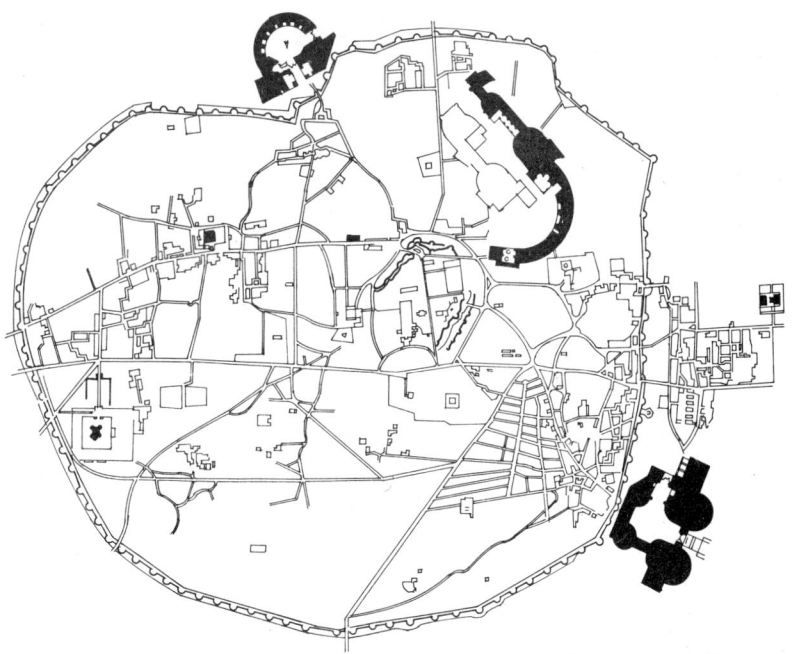

Fig 5.17 Plan of Bijapur city, AD 1550–1660

Adil Shahi kings to add suburbs like Shahpur on the north and Ainpur on east. It was only the suburbs of Nauraspur on the west, that at one time seem to have been connected to the citadel of Bijapur by the grand bazaar of Mohammed Shah. Yet, in spite of the apparent lack of the superficial and grandiose aspects of town planning, the necessary civic amenities were carefully provided for. As anywhere in hot and dusty India, it is the provision of water that is the most critical need of a living city. And in order that 'the thirsty lipped people of the world may drink to their hearts' content, Ali Adil Shah I had many extensive waterworks constructed. The city, of course, needed water not only for the 'thirsty lipped' but also for the 'Muslim nobility who loved to have it gurgling down his garden in ornamental channels,' and the Hindu populace for its ritual bathing. For this purpose, an extensive network of underground catchment tunnels and underground earthen pipes was laid to feed the various waterworks such as the Taj Bauri, the Chand Bauri *(Fig 5.16)* and Kumatgi. From these *bauris* which themselves became grand water pavilions and places of royal retreat, water was transported to the various parts of the city. Also, as in Ahmedabad, the *bauris* themselves became places of restful retreat and royal picnics. Of these, the more formally organized waterworks was at Kumatgi.

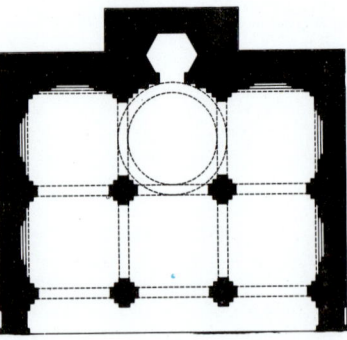

Fig 5.18a

Fig 5.18b

The Arq-Qila Citadel

Fig 5.18 *Malik Jahan's mosque at Bijapur, 17th century, (a) Plan, (b) View*

It was, whoever, within the walled enclosure of the original citadel popularly known as the Arq-Qila, that all famous buildings were located, such as the 'palaces and private apartments of the king and his family, various public buildings such as the civil and criminal courts, the military and revenue offices and treasury interspersed with courts of gardens, fountains, cisterns and running water.' Though the gardens and fountains have now vanished from the Arq-Qila, the prolific output of the Adil Shahis is reflected in the very statistics of the structures. There are as many as fifty examples of mosques and over a score each of tombs and palaces. These are to be found scattered at almost every town and corner of the streets of the city, covering

Fig 5.19a

Fig 5.19 Mecca Masjid at Bijapur, 17th century, (a) View, (b) Plan

an area of two-and-a-half square miles. It is obvious from the consistency of the architectural style that pervaded a century of intense building activity in Bijapur, that all these monuments were erected by a school of craftsmen that took the architecture of the Jami Masjid of Gulbarga and the tombs of the Barids at Bidar as its chief source of inspiration. From the former was borrowed the austere dignity of the stilted dome, the plain surfaces and the suavity of the arch, while from the latter came the bulbous or lotus dome which under the Bijapur builders became a thing of exotic beauty.

The Jami Masjid of Bijapur

The influence of the Gulbarga mosque is most apparent in Adil Shahi's Jami Masjid of Bijapur, erected some time in AD 1570. In fact, this building marks the true take-off point for the construction of the more significant architectural masterpieces of Bijapur. With the building of the Jami Masjid, Bijapur architecture had come a long way from the days of the early mosques of Karim-ud-din (AD 1320) and Khwaja Jahan (AD 1480), both constructed with despoiled Hindu temple material, and the coarse rubble, masonried and plastered three-arched and bulbous domed small mosques of Malik Jahan (*Fig 5.18*), the Mecca Masjid *(Fig 5.19)*. Ibrahim's old Jami Masjid and Ain-ul-Mulk of Ikhas Khan at Ainpur. All these earlier mosques varied only in minor architectural detail, and without exception consisted of three conjoined square chambers fronted by wide Gulbarga type arches with minarets planted in varying positions in the front facade. In the great Jami Masjid of Adil Shahi, though undoubtedly the idea of a covered central court was duly discarded and the slightly more fanciful ogee arch of the Gulbarga cloisters was transformed into a more stately and dignified arch of the four centre variety, the classic spirit of the prototype, nevertheless, pervades this mosque.

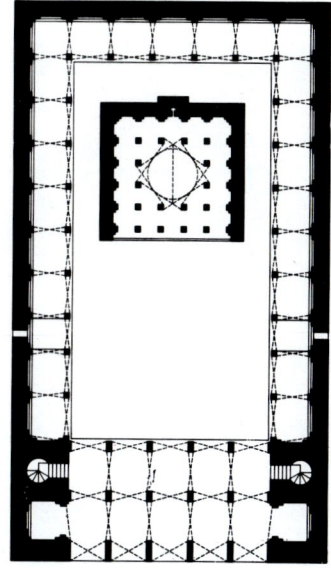

Fig 5.19b

The Friday mosque was never fully completed, its eastern wall and entrance gateway and minarets at the corners being left unfinished. But the *liwan*, measuring 208 ft × 107 ft (63 m × 33 m), a central dome rising up over the *mehrab* bay, and the unfinished northern and southern wings speak volumes of the architectural style *(Fig 5.20, 5.23)*. A single dome of the hemispherical variety covering the nine central bays rises over a richly ornamented square clerestorey platform, and at its apex is planted the familiar crescent finial symbolic of the Turkish origin of the Adil Shahi dynasty. The rest of the *liwan* is roofed over by shallow circular domes over square bays formed by masonry piers *(Figs 5.21, 5.22)*. These shallow domes seem intentionally concealed with the thickness of the roof. The seven stately arches of the facade of the *liwan* are shaded by a deep horizontal *chajja* supported over a row of closely spaced brackets. The routine exterior in most mosques would be formed of the large blank masonry surfaces of the walls surrounding the *liwan* and the side

Fig 5.20 The liwan facade of Jami Masjid, Bijapur

Figs 5.21, 5.22 Jami Masjid, Bijapur. Interior views of the shallow circular dome roofing over square bays, formed by masonry piers

Fig 5.21

Fig 5.22

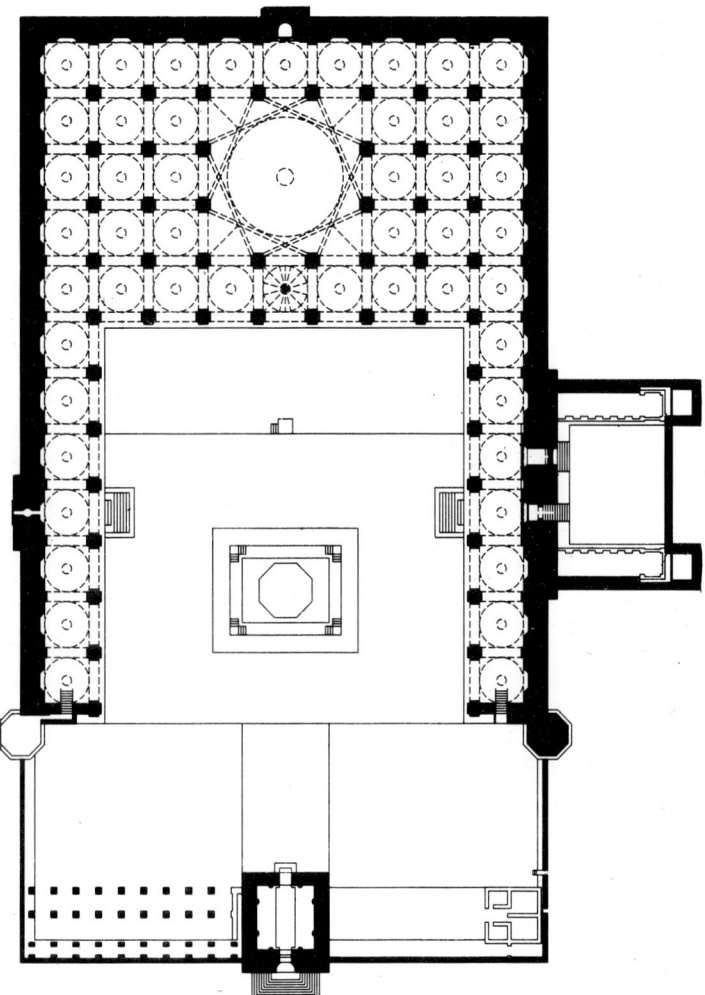

Fig 5.23 Plan of the Jami Masjid, Bijapur, AD 1570

wings. In this mosque these surfaces are cleverly contrived as a double row of deep arched niches, the lower one being blind, and the other admitting light and air into the *liwan*. This restrained but interesting treatment of the facade is reminiscent of the elevations of the *madrassa* of Mohammed Gawan at Bidar. It is, however, the interior of the *liwan* with its white plastered surfaces judiciously ornamented with deep grooved bands supplementing the robustness of masonry piers spanned by low imposted arches, that creates an impression of solemnity in this place of worship. The sacred grandeur of this bright white mosque is in many ways no less than that of the orange hued Jami Masjid of Mandu. In fact, in promoting the Islamic form of 'joint social worship of an intelligible God, the architects of these two mosques in India did all they could to keep the air flowing freely and light from the skies streaming in.'

The Tombs of Bijapur

In many tombs of Bijapur, roofed by onerous and at times oppressive domes, the light from the skies was prevented from reaching anywhere near the graves of the kings. The interiors are, in fact, so dark and gloomy that it is impossible for the eye to discern much without the aid of artificial light. For this, one cannot blame. Ali Adil Shah, the patron of the Jami Masjid. His love for open and airy structures is evident from the vast arched opening of the facade of the two-storeyed Gagan

Fig 5.24a

Mahal *(Figs 5.24a, b)* which was at one time his residence and council chamber. In his great humility, Ali Adil Shah had himself buried, too, in an open, airy, and almost veranda-like modest structure in the south-west of the city. In fact, the penchant for building these onerous tombs to perpetuate their own memory was given its earliest impetus under Ibrahim II (AD 1580-AD 1626) who erected his own burial complex in the form of a *rauza* containing not only his own tomb but a large mosque as well. This tradition had been alive, though in a minor key, in earlier Bijapur structures, and was probably borrowed from the Gujarat tradition where specimens of *rauzas* abound. Unlike the Gujarat examples, however, in the *rauza* of Ibrahim, it is the tomb structure that dominates the entire composition; the mosque merely provides the necessary architectural foil.

The Rauza of Ibrahim

The 115 ft (35 m) square structure of the tomb and the mosque stand at either end of a 360 ft (110 m) long and 150 ft (46 m) wide platform, located in the middle of a 450 ft (137 m) square sward of grass defined by inconsequential and hastily put together enclosing cloisters *(Figs 5.25)*. Some inking of the Mughal idea of a 'tomb-in-a-garden' seems to have inspired the Bijapur builders. The surrounding sward of grass, we are told, was once a grandly laid out formal garden. The grandest part of the entire scheme is, undoubtedly, the royal sepulchre. The central burial chamber of the mausoleum is of approximately 40 ft (12 m) side surrounded by two concentric arcaded verandas. The lower portion of the outer arcade is comparatively plain and rhythmically divided into stately arches of the Jami Masjid style. From the underside of the eaves of the deep *chajja* at the parapet level right up to the crescented apex of the bulbous dome, the facade is a crescendo of tier upon tier

Fig 5.24 (a, b) Views of the vast arched opening in the facade of the Gagan Mahal, Bijapur, AD 1560

Fig 5.24b

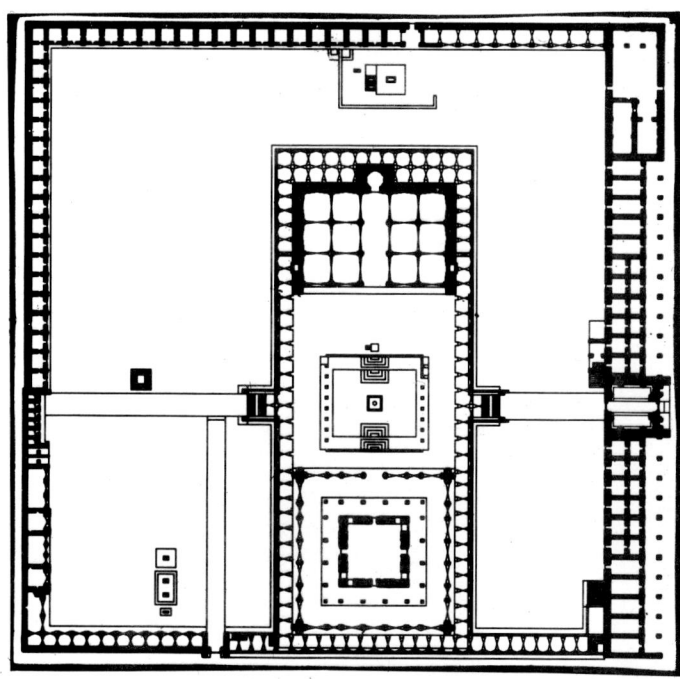

Fig 5.25 Plan of the Ibrahim Rauza, Bijapur

of richly embellished surfaces tautly strung together between the corner minarets *(Fig 5.26)*. The shafts of the minarets, with their rich horizontal mouldings, recall those of the mosque of Ahmedabad. That is, until one notices the spherical cupola resting with the petal leaf calyx at the apex, instead of the rather ungainly pyramid of the Gujarat style. The deep *chajja* with its multiple brackets and the richly battlemented parapet stretching from one minaret to the other is punctuated by short turrets that repeat the pulsating rhythm of the arches below. From behind and over this richly sculptured base and within the spatial frame of the minarets, rises the square stylobate of the inner chamber, also richly embellished and crowned by the familiar lotus dome set within a ring of large petal-shaped merlons.

Fig 5.26 Tomb in the Rauza of Ibrahim, Bijapur, AD 1615

Fig 5.27

Fig 5.27 The ingeniously crafted ceiling of the Ibrahim Rauza

Fig 5.28

Fig 5.28 Section of Ibrahim Rauza

Secret of the Hanging Ceiling

In order to raise the dome to a satisfactory external height, and yet not create a deep well of darkness in the interior, the 40 ft (12 m) square chamber is roofed over at an intermediate level by an ingeniously crafting ceiling *(Figs 5.27, 5.28).* The flat portion measures 24 ft (7.3 m) square and virtually floats over 7 ft (2 m) deep brackets projecting from the walls around. The only explanation preferred for this unbeamed and seemingly unsupported ceiling not having fallen down in the possible use of some secret formula for the rich mix of the mortar by the Bijapur builders. This mix not only enabled the area to be spanned but also kept in place the underlayer of stones which are merely butt jointed with each other. Whatever be the clue to this miraculous construction, there is little doubt that the architectural style of the tomb and its complementary mosque which echoes the volumes and forms of the tomb, arouses strong mixed emotions. While this group is said to be 'as rich and picturesque as any in India and far excelling anything of the sort this side of the Hellespont,' and 'every part, whether structural, technical, ornamental or merely utilitarian appears to have been thought out and provided for in a most meticulous manner,' at the same time 'they make up a group of gorgeous but, it must be confessed, somewhat barbaric splendour.'

Another eye-catching composition, but of somewhat muted splendour, is the rather curiously named Mehtar Mahal, which is an entrance gateway to a small mosque. In this small but tall facade *(Fig 5.29)* the Bijapur builders truly went to town in decorating it not only with flat geometric patterns but fanciful brackets, deep balconies and sloping *chajjas* all framed within the two slender minarets. With the construction of the Ibrahim Rauza and the Mehtar Mahal, their seemingly insatiable thirst for enrichment of detail appears to have been finally quenched. The Bijapur craftsmen, too, like the Hindu builders of the temples of Halebid and Belur had reached saturation point in creating profusely decorated surfaces. But unlike the Hindu, they had within their bag of tricks one more of which they had shown glimpses but which had not yet been fully tapped by their Adil Shahi patrons. Sultan Mohammed (AD 1626) astutely realized that in richness of embellishment the Ibrahim Rauza was not to be outclassed, and decided to built a tomb for himself that could not be outdone in sheer massiveness and size. It is to the great credit of his builders that his ambitions were more than fulfilled. They built for him a tomb that was roofed by a dome that came to be rated as the largest anywhere in the world — the famous Gol Gumbaz of Bijapur.

Fig 5.29 Fanciful brackets, deep balconies and sloping chajjas of the Mehtar Mahal, Bijapur, AD 1620

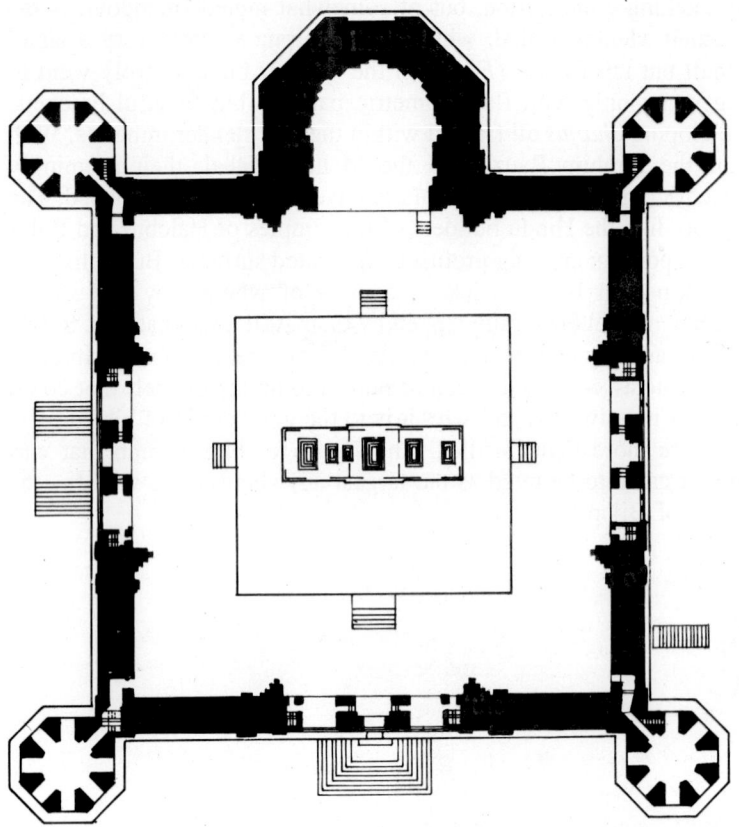

Fig 5.30 Plan of the Gol Gumbaz, Bijapur, AD 1656–60

Statistics of the Gol Gumbaz

Transcending all other buildings at Bijapur in its volume and mass, this tomb of Sultan Mohammed commenced construction in AD 1656, and its rubble masonry walls were still being plastered when the ruler died. That immensity of size was the major criterion for erecting this tomb is apparent from the building plan *(Fig 5.30)* which is simply 'a square hall enclosed by four lofty walls, buttressed by octagonal towers at the corner, and the whole surmounted by a hemispherical dome' *(Figs 5.31, 5.32)*. The bare vital measurements of the entire structure are rather startling. For the 'simple square' is almost of 136 ft (41.5 m) side inside and as much as 205 ft (62.5 m) outside; the four lofty walls are over 10 ft (3 m) thick and 110 ft (33.5 m) high; the diameter of octagonal buttresses is 25 ft (7.6 m) rising to a height of 150 ft (45.7 m); the hemispherical dome is of 144 ft (44 m) diameter outside and 125 ft (38 m) diameter inside; its apex is over 200 ft (60.9 m) from ground level. The whole structure in height alone is thus the equivalent of a 20-storeyed structure of modern times.

Method of Intersecting Arches

It would be apparent then from these statistics, a study of the plan and rather plain exterior of the monument, that the moving spirit behind this great building venture was not a conventionally trained architect but a daring structural engineer. It is indeed remarkable that without precedent to prove his skill, the master builder (probably one Yaqut of Dabul) could persuade Sultan Mohammed that he would be able to successfully build for him a monument of such gigantic proportions. The crux of the whole design, as described earlier, became that of supporting the circular dome over the cube below and intelligibly managing the phase of transition from

Fig 5.31 *External view of the Gol Gumbaz*

the cube to the dome above. Earlier this had been managed by a lintel, a series of pendentive or squinch arches across the corner of the cube below depending on the size of the Gol Gumbaz, none of these methods could be applied. The span across the corners of the square itself would be of the order of 75 ft (24 m), and by this conventional method the diameter and consequently the load of the dome would have been twice that of the present dome. Thus, the problem was to somehow reduce the size of the dome while retaining the huge size of the square hall below. This was solved by the builders by employing what has come to be known as the 'method of intersecting arches.' This procedure had been earlier adopted for erecting the dome over the Jami Masjid at Bijapur. But in the Jami Masjid, it seemed to have been used to produce a pleasing interior composition, while in the case of the Gol Gumbaz, it was a sheer structural necessity.

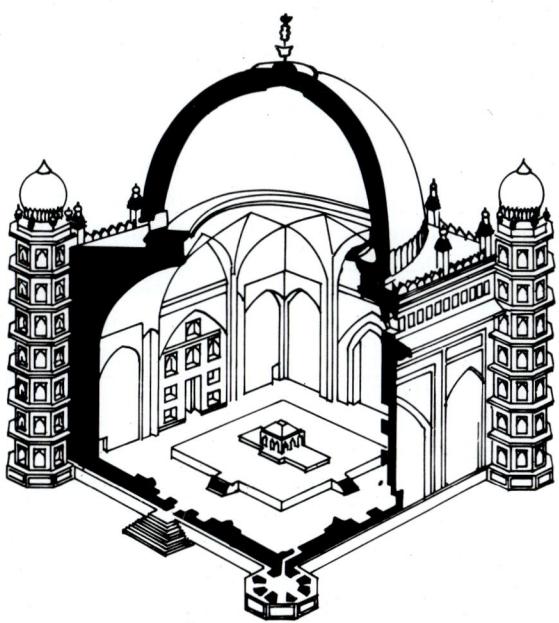

Fig 5.32 *Isometric view of the Gol Gumbaz*

The geometric essence of this solution lay in inscribing within the large square, two smaller overlapping squares, by dividing each side of the large square into three equal divisions and joining together the alternate points of division(*Fig 5.30*). As can be seen, eight points of intersection of the two smaller squares produced an octagon within the large square of a size smaller than the octagon produced merely by chamfering the corners of the square. Continuing with the geometric analysis, it is the octagon that could now be gradually made to approximate to the required circular plan of the dome. Structurally, the location of the eight corners of the octagon in space was determined by erecting tall arches along each of the sides of the intersecting squares *(Fig 5.33)*. The points at which these arches intersected in the volume above the large square became the corners of the octagonal platform over which a circular ring of masonry acting as the drum for the dome could be erected.

The Largest Dome in the World

The hemispherical dome of an average thickness of a much as 10 ft (3 m) at its springing point was constructed out of concentric oversailing layers of brick masonry cast in concrete formed out of a mix of ballast and rich lime mortar. Such a homogeneous shell or monoblock of virtually brick 'reinforced' concrete was constructed without the use of any scaffolding or shuttering of timber, except for the section near the crown.

In principle, the construction of this dome differed vastly from the Roman, Byzantine or Renaissance dome. In the Roman and Byzantine domes, a massive load had to be appended on the outer surfaces of the haunches of the domes to restrain the corresponding thrust of the dome. The Renaissance builders, on the other hand, used cumbersome rings of iron chains to surround the drum of the dome. In the Gol Gumbaz, this restraint was provided by the masonry poised over the intersecting arches within the dome. This, while the dome of the Gol Gumbaz can be seen externally in its full splendour, that of the Pantheon at Rome, though larger in diameter, makes hardly any external impact, its lower circular form being hidden behind a stepped-back mass of masonry. True, the dome of the Pantheon and that of St Peters are both larger than the Gol Gumbaz in diameter. However, the former two rise vertically over a circular plan of the same size as the diameter of the dome. The Bijapur dome, on the other hand, which is poised over a larger square space below, spans the largest uninterrupted floor space in the world, of the order of more than 18,000 sq ft (1,672 m).

Having said so much in praise of its structural achievement, it must now be pointed out that apart from the whispering gallery fortuitously formed at the balcony level above the apex of the intersecting arches (since the circumference of the dome happens to be just right to produce a double echo), the Gol Gumbaz has little else of interest or architectonic excellence. Because of its immense size, the outer surfaces proved much too vast to be organized into a pleasing architectural composition. Thus, three of the outer walls are composed of arches filled in completely with screens formed out of the dull brown local basalt. The fat, octagonal seven-storeyed turrets, too, borrowed most likely from the Charminar of Hyderabad, contribute little to the elegance of the elevation, apart from being structurally redundant. The vastness of the interior, too, which one enters rather abruptly through the thickness of the outer wall, could have been brought to life by some form of controlled natural lighting. As it is, in the darkness it is difficult even to comprehend the vastness and majesty of the space enclosed by the gigantic dome. Thus, while the Bijapur builders had erected the largest ever tomb in India, it was left to the succeeding Mughal builders of Delhi to create the most elegant one, the Taj Mahal at Agra.

The stupendous effort of building the largest dome in the world appeared to have exhausted the creative energies of the Bijapur builders, as well as the strength

Fig 5.33 Section of the Gol Gumbaz

of the Empire. It must, however, be said to the builders' credit that their ambitions remained undaunted, though these were duly tempered by the political situation prevailing at that time. Both the Mughals and the Marathas had been making little dents into the power of the Adil Shahis at Bijapur. And thus, though the foundation of a building envisaged to be as large as the Gol Gumbaz and identical configuration were laid out at Ainpur as a tomb for Jahan Begum, a smaller chamber was introduced within. This would have avoided the necessity of building yet another great dome. However, the builders did not even get beyond the lintel level of this structure. Likewise, the tomb of Ali Adil Shah II, based on the plan of Ibrahim Rauza, did not go beyond the erection of the soffits of the arches of the double arcade around the central square tomb.

After the death of Ali Adil Shahi II in AD 1672 and the accession of his son Sikander, not much remains to be told of the Bijapur story. 'The falcon was hard upon its prey and was about to make a final swoop' which he (Aurangzeb) did in AD 1686 six years after Shivaji, the Maratha tormentor of the Adil Shahis had died. Aurangzeb lived in the city of Bijapur for some years after its fall. Finally, in AD 1724, it was ceded to the Maratha Peshwa, who found in its public buildings a 'mine of materials.' The palaces 'were stripped of all their woodwork; beams, doors, windows were ruthlessly torn out and carted away.' But, as we have discovered in reviewing the architectural remains of the Bijapur builders' efforts, the spirit of their enterprise could not be 'carted away.' Even in its derelict splendour, the remains of the handiwork of the shortlived Adil Shahi dynasty of Bijapur point to the task and the high power of art and erection which they evince.'

*Bare Khan ka
Gumbad, Lodi
Gardens*

CHAPTER 6

The Lodis, Mughals and Sher Shah

AD 1414–AD 1560

After the violent upheaval caused in Delhi by Timur in AD 1398, architectural activity had been considerably subdued in the capital for a long period of time. It is at this juncture of inactivity and transition that a survey of building in the provinces of India has been made. Returning once again to Delhi, we find that in the early fifteenth century architectural activity may have been stilled, but political intrigues, conspiracies and murders were rife. The courtiers and Amirs of the defunct Delhi court raised a Lodi noble to the throne. Daulat Khan, the Lodi, was soon unceremoniously removed by Khizar Khan, the Governor of Multan in AD 1414. Khizar Khan, who claimed dubious descent from the Prophet's family and continued to owe allegiance to Timur, styled his dynasty as that of the Sayyids. His various successors, most of whom were 'victims of factions and sport of circumstances,' ruled the ever shrinking so-called Kingdom of Delhi with wretched ineptness. The 'kingdom' was, in effect, ultimately reduced to that of the city of Delhi and a few neighbouring villages under the rule of the rather pretentiously styled Ala-ud-din Alam Shah in the year AD 1451. Prudently, the last Sayyid, Alam Shah, retired to a life of pleasure in his favourite Badaun, leaving Buhlul Lodi, chieftain of an Afghan tribe, to take over the reins of power. Buhlul was the first true blooded Afghan to sit on the throne of Delhi, after centuries of rivalry with rulers of Turkish descent.

Sayyid and Lodi Necropolis in Delhi

The power, or rather the lack of it, of the rulers of Delhi at this stage was so acute that even provincial rulers such as the Sharqis of Jaunpur had the temerity to attack the capital city. However, Buhlul was made of somewhat sterner stuff than his Sayyid predecessors. He was possessed of sufficient courage, energy and tact to be able to restore some vestige of Muslim prestige in Delhi before the time of his death in AD 1489. He was followed by Sikander Shah who has been highly praised by contemporary writers for his excellent 'qualities of head and heart.' It was under his rule that in a way the 'head' or the centre of government of the Lodis was shifted to Agra, and only their heart remained in the old capital. While the Lodis ruled from Agra during this period, their mortal remains were invariably brought to Delhi for burial. All the more prominent architectural monuments of the Lodis, largely consisting of their tombs, are to be found in Delhi. In fact, the output of the tomb builders was no prolific that Delhi and tracts around it became one vast necropolis. More than fifty tombs of kings and ministers and other members of the Afghan nobility were erected in the countryside south of Firuz Shah's city of Tughlaqabad.

Tombs in the Lodi Gardens

The most impressive group of these tombs, that of the kings and his ministers, is to be found today in the heart of the modern city of New Delhi in what has come to be popularly known as the Lodi Gardens, formerly the Lady Willingdon Park.

It is apparent that two distinct prototypes of the tomb design were developed: one that was octagonal in its configuration, and the other, more conventional, consisting of the familiar cube with a domed roof. It was the octagonal composition that caught the fancy of royalty—probably because of its association with the plan of the sacred Dome of the Rock at Jerusalem. However, the immediate inspiration for this type of structure lay closer to home; in fact, just a mile or so east in the holy retreat of Nizam-ud-din Auliya. As we have already seen, the first octagonal tomb in India had been erected over the remains of Khan-e-Jahan Telengani, in the last days of Tughlaq rule. In building a tomb for Mubarak Shah in AD 1434, the Sayyid builders proceeded to enlarge and refine the proportions of the Tughlaq prototype *(Fig 6.01)*.

Fig 6.01 Tomb of Mubarak Shah, New Delhi, AD 1424

Each of the distinct elements of the tomb, the 50 ft (15 m) high dome, the arches in each of the 30 ft (9 m) sides of the octagon, the merlons along the parapet and the kiosks above the verandas are all more firmly organized than in the Tughlaq prototype. Yet, the stilted dome seems to sit rather squatly over its 70 ft (21 m) wide octagonal base. This fault was, however, duly rectified in the tomb of Muhammed Sayyid built just a decade later. Here, the drum of the dome along with the ornamental kiosks around it was raised several feet higher. This critical rectification resulted in the most handsome example of Sayyid architecture in Delhi *(Fig 6.02)*. The finely stilted dome of the tomb is now gracefully poised over a substantial base rather than awkwardly pressing down upon it.

Fig 6.02 Tomb of Muhammed Sayyid, Delhi, 1444

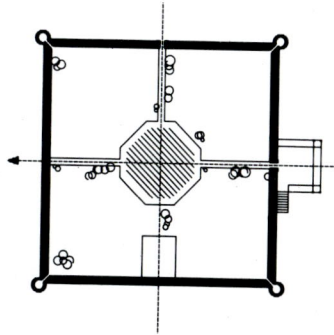

Fig 6.03 Plan of tomb of Sikander Lodi, New Delhi, AD 1518

The Garden Tomb of Sikander Lodi

The profile of the tomb of Mubarak Sayyid was found to be so pleasing that when seventy-five years later in AD 1518 a tomb had to be erected over the remains of Sikander Lodi, almost an exact copy of the Sayyid tomb was repeated *(Fig 6.03)*. Only the kiosks on the veranda roof were eliminated and replaced by semi-minarets or *guldastas* attached to the base of the tomb *(Fig 6.04)*. However, the Lodi builders, in terms of structure and overall layout made two significant contributions to the development of Islamic architecure in Delhi. First, in an attempt to raise the outer dome to even more splendid heights, without allowing the inner chamber to appear disproportionately tall, they introduced a second inner dome of a lower profile within the outer one. By building two domes in the form of an inner and outer shell separated by a void, the proportions of both the exterior, and particularly the interior, were much improved. Second, though the structure is not quite set within an ornamental garden — as later Mughal tombs were — for the first time in Islamic Delhi, the tomb, instead of standing forlorn in barren and flat country, became the focus of a fairly formal and elaborate arrangement. A deep and arcaded surrounding well defines the square courtyard in the centre of which lies the tomb. By the addition of a veranda to its western wall, the central became the *liwan* and mosque attached to the funerary chamber. Impressive gateways, too, were installed at the cardinal points in centres of the surrounding walls. The one in the southern wall with a large platform is particularly elaborate in its arrangement. Although the surrounding wall is still fairly fortress-like in character, the tomb layout marks a definite point in the progress made from the rather elementary 'tomb and mosque' of Sultan Ghari. At the same time it anticipates, though only in an embryonic form, all the paraphernalia that characterized the later grand 'garden and tomb' of the Mughals.

Fig 6.04 *View of Sikander Lodi's tomb in the Lodi Gardens, New Delhi*

'Gumbads' for the Lodis

For other non-blue-blooded dignitaries of the Afghan court, on the contrary, tombs of the more conventional 'cube and hemisphere' type were being profusely built all over the growing 'grand cemetery' of South Delhi. The large and impressive boat keel profiled domes crowning these various tombs are the most outstanding architectural features of these monuments. So much so, that in local parlance, all these tomb structures came to be designated merely as *gumbads* (domes). And we have a long list of these. The most prominent ones are the Bare-Khan-ka-Gumbad (AD 1497), Chote-Khan-ka-Gumbad (both in the modern South Extension area of New Delhi), and the Bara Gumbad (AD 1494) and the Shish Gumbad in the Lodi Gardens. The design basis of these tombs is a sort of uninspired amalgam of various architectural ideas that were then afloat in and around Delhi. The plan formulated to build the thirteenth century gateway or Alai Darwaza of the Qutb mosque, was the basic format followed by the Lodi builders, an example of which is the Chote-ka-Gumbad *(Fig 6.07)*. On to the blind arched facade of this prototype, they grafted a rather subdued version of the central arched pylon of the Tughlaq or Jaunpur variety, devoid of any inclined surfaces. At the bottom of this fronton was installed a modestly proportioned bracket and beam entrance doorway of distinctly Hindu origin. The tall vertical surfaces of the walls of the cube below — again in the Alai Darwaza style — are broken up into two or three 'storeys' defined by a series of blind arches. Some of the tombs, such as the Bara Gumbad in the modern Lodi Gardens *(Fig 6.05)*, show signs of having been at one time decorated with coloured glazed tiles. In their present bereft stage, though awesome in size — the Bare-Khan-ka-Gumbad *(Fig 6.06)* rises to as much as 80 ft (24 m) from the ground — these innumerable tombs of the Sayyids and Lodis do not exhibit any significant aesthetic development in the architecture of the Indian Islamic tomb. However, the rather bare surfaces of these are devoid of the normal decorative trapping of the Hindu architectural style. The robust confidence with which the large true Lodi domes have been structured, undoubtedly marks the gradual ascendancy of Islamic engineering techniques over the sculptural architectonics derived from indigenous sources.

The Jamala Masjid and Moth-ki-Masjid

The ascendancy of Islamic concepts is also clearly apparent in the moderate sized mosques of the Lodis. Even in their uncertainty of design they pave the way for the arrival of the great mosques of the Mughals. Until now the Tughlaq builders had been 'screening' the facades of the *liwan* of their mosques with a turreted pylon or pylons. The Lodis decided that their bold arched facades and boat keel domes needed no such artificial aid *(Fig 6.09)*. And so the Delhi mosques of the Lodis, such as the Moth-ki-Masjid (AD 1505) *(Fig 6.08)* and the Jamala Masjid (AD 1536) *(Fig 6.10)*, are a clearer and more honest expression of the Islamic arcuate style than those of their Tughlaq predecessors. But again, as in their tomb architecture, the Lodi builders were

Fig 6.05 View of Bare Khan ka Gumbad, Delhi, AD 1497

unable to impart any great delicacy to the undoubtedly rather bold but bare expressions of the arch in the facade of the *liwan*.

The depth of the wide, boat keel-shaped arch in the Lodi buildings is broken up into multiple flat receding panels rather than mouldings, as in this mosque attached to the Bara Gumbad in the Lodi Gardens. In fact, the very profile that emphasizes strength when used in the dome represents an uncertainty and weakness when applied to the arch. Apart from the lack of mere decorative ideas, the builders of these mosques did not have a comprehensive control over their architectural formulations. This is apparent from the fact that the rear and the front facades of the *liwan* of the Jamala Masjid are derived from two completely different ideals. Whereas the design of the rear wall of the *liwan* with its prominent circular buttressed turrets is certainly a hangover from the Tughlaq period, the front is avowedly contemporary and Lodi. In a way, it seems that all that the builders did was to relegate the front elevation of the Tughlaq mosque to the rear of their certain and merely suggest a new facade for the *liwan*.

Fig 6.06

Fig 6.07

Fig 6.06 Bare-Khan-ka-Gumbad

Fig 6.07 Chote-Khan-ka-Gumbad, Delhi

Fig 6.08 Moth-ki-Masjid, Delhi, AD 1505

Babur Invades India

For all these raw but robust architectural ideas to find more graceful expression, it needed not only many more years but also momentous changes in the Islamic body politic of India. The architecture of any civilization is the true and inviolate barometer of the state of health of that civilization. It would be apparent from examples of Lodi building enterprises that Delhi, though abounding in wealth of craftsmanship, was crying aloud for fresh design and political impulses. In spite of going through the trauma of yet another foreign invasion, it was indeed fortunate for Islamic India that this invasion was spearheaded by none other than Babur, the founder of the Mughal dynasty. In AD 1526 (well before the Jamala Masjid was complete), on the strategic fields of Panipat with only 10,000 men, Babur defeated the ominous, 1,00,000 men army of the Lodis and slew the last Lodi ruler, Sultan Ibrahim. Apart from being a

Fig 6.09 *Elevation of a typical Lodi mosque, Delhi, 15th century*

Fig 6.10 *View of Jamala Masjid, Delhi, AD 1536*

military genius, Babur had also been endowed with a keen artistic temperament, a quality that was fortunately inherited by the majority of his descendants.

It is true that Babur himself did not leave any concrete and direct impact on the course of Islamic architecture in India. But his grand pronouncements in his famous memoirs, the *Babur Namah*, on the quality of the local architecture and other traditions, played a vital role. He wrote in these memoirs that, 'Hindustan is a country that has few pleasures to recommend it... The people have... no ingenuity or mechanical invention in planning or executing their handicraft works, no skill or knowledge in design or architecture... no grapes or musk melons, no good fruits, no ice or cold water... There is an excessive quality of earth and dust flying about. But a convenience of Hindustan is that the workmen of every profession and trade are innumerable and without end.' In his criticism of India it can be easily read that Babur was languishing for his native Kabul with its flowing water channels amidst fruit and flower laden gardens. No wonder then, that as soon as he got respite from his numerous campaigns in India and retired to rest in Agra, his first act of building — if it may be called one — was that of laying out a formal garden *(Fig 6.11)*. In a country that according to him lacked also the quality of 'symmetry in architecture,' he ensured that the garden would be planned out on strictly geometric and symmetrical patterns. It was his early pioneering efforts that resulted in the fabulous tradition of the now famous Mughal gardens of India. Some excavated remains of the earliest of these have been discovered on the outskirts of Dholpur near Agra.

Fig 6.11 Building in progress for Babur at Agra

Indian Masons Build for Babur

In the field of building, it is recorded that Babur called for the pupils of the famous architect Sinai from Constantinople to execute his schemes in India. At one time, '680 Indian masons worked daily on his buildings at Agra, and 1,500 were daily employed on his buildings at Sikri, Biana and Gwalior.' The only identifiable and extant example of his work is a *masjid* at Panipat. Except that it was largely erected in brick it does not exhibit any order distinct quality.

Babur' earlier life of thrilling adventure and hardship while roving in Hindustan and in the rather forbidding mountains around his home of Ferghana, finally caught up with him after he had been a mere four years in India. He died at the comparatively young age of forty-seven years. As he himself candidly records, his had been a life packed with as much success as adversity. Wild adventure, hard endurance, literature and poetry, homosexuality, drinking and drug addiction and grand military and political exploits were the ingredients of his heroic life. The hallmark of Babur's genius was an essentially liberal bent of mind. Bigotry could never be attributed to him. It was this quality of tolerance that Akbar, his grandson, imbibed to the fullest, and which helped in his consolidating the great Mughal empire. Babur's hectic political life which had begun at the tender age of eleven, was a legendary one, and so was his death. His eldest son, Humayun, was virtually on his death bed when his courtiers suggested that the famous diamond Koh-i-noor be placed as a 'meet pledge' for his life. But Babur, with his characteristic sincerity determined to place his own life in the bargain. 'Oh God, if a life may be exchanged for a life, I, who am Babur, give my life and being for Humayun.' It is said that he soon felt a fever grip him and he was in ecstasy on having taken on Humayun's dangerous illness.

Humayun and Sher Shah

Babur forfeited his life in the final act of sacrifice on 26 October 1530, while Humayun recovered fully. Thus, Babur's journey from the snows of Turkestan to the sunlit banks of the Yamuna was finally concluded. Humayun's own journeys which were to carry him through a tortuous but fruitful exile from India were just commencing. Even though, in a way Babur, with his great victory over the combined Rajput forces under Rana Sanga at Khandwa, had laid the foundations of Mughal superiority in northern India, Humayun, the indulgent, was unable to quite hold on to his empire. Architecturally speaking, too, after fine astrological calculations, in the year AD 1533, Humayun laid the foundations of a new city at Delhi, ambitiously styled Din Panash or Asylum of Faith. The walls of the city were, however, not destined to provide shelter to Humayun for too long. He was able to erect the rubble masonry boundaries and some sort of makeshift palaces for himself on his new, elevated site that was then almost an island in the Yamuna. Referred to as Indraprastha by historians who associate the site with that of the legendary Pandava city, it is today popularly known as the Purana Quila or Old Fort. Before the city could become an impregnable fortress, it was usurped, not by the utterly disarrayed Rajput forces, but by an extremely able and enterprising Afghan ruler based in Sasaram, Bihar, not far from the eastern Sharqi kingdom of Jaunpur.

Fig 6.12 Lal Darwaza, opposite Purana Qila, Delhi, 16th century

Sher Khan or Sher Shah Sur, had established himself as the leader of the many Afghans who, over the years, had settled along the Ganges in Bihar. He had paid due allegiance to the ruling Babur. But finding in Humayun a weak ruler, by a clever mixture of political and military strategy, he twice convincingly outfoxed the Mughal armies — once at Chausa in Bihar, and again, more decisively at Kanauj in May 1540. On both occasions Humayun had to retreat in disarray across the Ganges. The second defeat was destined to be long and bitter. Sher Shah Sur gave no respite to Humayun, and his own brothers Hindal and Kamran extended him no refuge in their provinces of Kandahar in Afghanistan, Humayun, left with no choice, fled to Persia in AD 1544, leaving his fourteen month old infant son, Akbar, in his brother Kamran's care. It took Humayun more than a decade of virtually wandering from door to door before he was able to recapture Delhi. Humayun's fortunes may well have been on the wane, but not those of Delhi and northern India. After the inept rule of the Sayyids and Lodis, in a brief period of just three decades, northern India was ruled, in quick succession, by three most magnificent personalities — Babur the Mughal, Sher Shah the Afghan, and Akbar the Great. Sher Shah was now busy in Delhi laying the foundations of social, administrative and architectural reforms which Akbar later adopted in building his great empire.

Fig 6.13 Khuni Darwaza near Tughlaqabad Gate, Delhi, 16th century

Delhi Sher Shahi

On taking over the fortress of Humayun at Delhi in the true tradition of Indian Muslim rulers, Sher Shah proceeded to lay out a new city that he intended to call Shergarh but that came to be popularly known as Delhi Sher Shahi. The city, laid out east and north of the existing fortress, was envisaged to have a surrounding city wall more than nine miles (14 km) long. Some remains of this wall have been found as far north as the present Ajmer Gate of Shahjahanabad. However, apart from the so-called Lal Darwaza (*Fig 6.12*) located directly opposite the eastern gateways of Humayun's Purana Qila, and the so-called Khuni Darwaza (*Fig 6.13*) east of the Tughlaqabad gate (on the present Mathura Road), nothing much is visible today of

Fig 6.14

Sher Shah's city of Delhi. In fact, the limits of Shergarh are a matter of conjecture and dispute. Some remains, believed to be of Sher Shah's city, were discovered during excavations carried out for the building of Delhi's new High Court near Lal Darwaza. The Archaeological Survey of India, however, did not consider these to be worth preserving. Though the exact location of Delhi Sher Shahi is unknown, the robustness of the architecture of the majestic gates of the city is indicative of a building tradition coming to life again under a new and powerful guiding hand. All the inconsistencies of the profiles of the Lodi arch are duly resolved into extremely clear delineated and handsome arches. What is more, even though marble and coloured tiles are sparsely used, these were obviously applied by the hands of extremely competent craftsmen.

Fig 6.15

Fig 6.14 Interior of Qila Kunha Masjid, Delhi

Fig 6.15 Qila Kunha Masjid within Old Fort, Delhi, AD 1545

The Sher Mandal and Qila Kunha Masjid

The true competence of the craftsmen of Sher Shah is proven more by what may well be considered the watermark of pre-Mughal mosque design — that of the Qila Kunha Masjid within the Old Fort at Delhi *(Figs 6.14, 6.15)*. During the time of Sher Shah, it appears that Humayun's Purana Qila had become the Governor's residence, and it is within this that the chapel royal was erected. True that there are few fortresses in India which recall more vividly 'the days of medieval pageantry and oriental ceremony than Sher Shah's citadel,' but the only two structures standing within its vast emptiness are the mosque and the so-called Sher Mandal. On the one hand the octagonal shaped three storeyed pavilion of the Sher Mandal *(Fig 6.16)* carries an air of mystery about it (the function of the structure is obscure to date). The mosque, on the other hand, has a convincing air of architectural mastery. Though only the sanctuary measuring 158 ft × 45 ft (48 m × 14 m) and rising to a height of 66 ft (20 m) is standing today,

Fig 6.16 Sher Mandal within Old Fort, Delhi, 16th century

its front facade in its delicately balanced composition reveals that the revival of the building arts in Delhi had reached a stage of consummation. In its overall architectural scheme, the builders accepted the five archways and a single dome composition of the Jamala Masjid as a model. They eschewed, however, the ungainly ogee multiple recessed arches of the prototype and replaced them with four central, firmly contoured Tudor arches. Of varying height, and some with the spear profile, these are installed within well-defined rectangular frames. The builders of Sher Shah had thus added a certain stateliness to the rather amorphous and weak composition of the mosque. What is more, there is a certain vibrancy in the play of light and shadow in the *liwan* facade, created by the variety of means used to frame one smaller arch within another, each in a different plane. The same lively approach to the building arts is evinced in the different methods used to effect the 'phase of transition' from the square bay to the dome. While in the central bay the traditional squinch arch has been employed, in the next bay a very attractive variety of stalactites is used, and in the last, a semi-vault of unusual design. The Sher Shah builders, too, proved true to the idiom that the Afghans 'built like giants and finished like goldsmiths.' The *mehrab* in each of the archways with its decorative treatment of the imposts, the rich foliation of the major arch, and the delicately inscribed border is a notable illustration of applied art. It is once again only in the rear facade of the *liwan* of the Qila Kunha Masjid that the builders were unable to shake off the Tughlaqian tradition. Such features as circular turreted buttresses are duly planted at the corners as well as behind the central portal. In fact, it would seem that the militant notes of Tughlaq architecture had become a part and parcel of subsequent Islamic architecture in Delhi. Some of these figures appear prominently even in the works of the Mughals.

The Great North-Eastern Highway

The Lodis had shifted the centre of government to Agra, and Delhi became their burial ground. On the other hand, for Sher Shah, Delhi became the 'head' but his heart remained in his hometown of Sasaram. As if to link the head and heart with a living artery Sher Shah made the Herculean civil engineering effort of laying a road to connect the eastern provinces of India with Delhi and even further north. The present Grand Trunk Road connecting Delhi with Calcutta follows almost the same alignment as Sher Shah's grand north-eastern highway. Probably due to creation of this link along which an efficient postal system was established, the importance of Sasaram was maintained. Even before Sher Shah became a national instead of provincial ruler, his eyes were riveted to Delhi. This is proven by the fact that in building the tomb of Hasan Khan, his patron's father, he chose the Lodi prototype of Delhi as his model. In this initial effort at aping the Delhi style, his architect, one Aliwal Khan, somehow fumbled through and managed to produce a fair but undistinguished replica of the Lodi tomb. When Sher Shah died during an expedition in Kalinga in AD 1545 after a brief but dazzlingly brilliant rule of just over five years, he was buried in Sasaram. In the building of a tomb for his king, the architect Aliwal, having mastered the technique of the octagonal tomb, took the design of this type of mausoleum to its climax, both in terms of proportions and size.

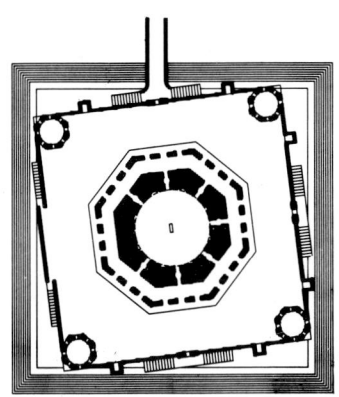

Fig 6.17a

Fig 6.17 Tomb of Sher Shah, Sasaram, Bihar, 16th century, (a) Plan, (b) View

Fig 6.17b

Sher Shah's Tomb at Sasaram

The magnificent tomb of Sher Shah, standing in the middle of a large square sheet of water of 1,400 ft (426.7 m) side is approached through a gateway built on the mainland that is connected to the tomb by a causeway on its northern side *(Fig 17a)*. The tomb itself became an 'immense pyramidical pile of ordered masonry in fine distinct stages,' the whole rising to a considerable height of 150 ft (45.6 m) and having a square base of 250 ft (76 m) side. The main tomb in its essential configurations is not strikingly different from its Lodi prototype except in terms of size. What adds immensely, though, to its architectonics is its site and location. Planted as it was in the midst of a large waterbody, it had to be erected over an even more spread-out low stylobate with steps gracefully leading down to the level *(Fig 6.17b)*. As is apparent, an error in orienting the tomb along the cardinal axis was corrected after this lower platform had been built, resulting in the curious diagonal relationship between the lower and upper plans. It is the domed canopies planted at each corner of the square platform, and the familiar kiosks over the base, as well as the dome rising up from over a tall octagonal drum that create the silhouette of a 'five storey pyramidical mass rising out of the waters.'

Picturesque Salim Garh

The octagonal type of tomb had reached the acme of its architectural appeal with the building of Sher Shah's tomb. Nevertheless, repeated attempts were made to rival this monument, but in vain. Isa Khan, a minister of Sher Shah's successor and son of Salim Shah, built himself a tomb near the present more well-known Humayun's Tomb. This, however, turned out to be a replica of the tomb of Sikander Lodi, complete with a complementary mosque and courtyard. Salim Shah himself ventured to build a grander version of his father's tomb in Sasaram. This was never quite completed after his early death. But Salim Shah did manage before his death to build a rather picturesque little fort in Delhi in AD 1546. Well north of the Old Fort and Tughlaqabad and located on an island in the Yamuna, it came to be known as Salimgarh. This fort was later connected by a bridge to Shahjahan's Red Fort, presumably as a prison. It is today occupied by the Indian Army and is unapproachable to visitors. Moreover, the East Indian Railway laid a track cutting across the fortress, and no buildings of any historic consequence survive within.

The Return of the Mughals

Salim Shah himself had apparently lost interest in the city of Delhi and moved his capital to the city of Gwalior. His death in AD 1554 was the signal for the collapse of the Afghan house. Civil war ensued between Sikander Sur, the Governor of Punjab, and Muhammed Adil Shah, the proclaimed successor of Salim Shah. The latter's Hindu minister and general, Hemu, seized northern Delhi and proclaimed himself Raja Vikramaditya. In the meanwhile, after fifteen years of exile, Humayun seized the opportunity to recapture his lost territory. He succeeded in entering the Purana Qila in July 1555 and remounted his father's throne. In spite of being surrounded by hostile Afghan forces, Humayun was able to pass his last few days in his favourite pursuit of reading poetry while his generals fought his wars for him. In fact, poetry was almost the lingua franca of the newly established Mughal life. These poetry sessions were most likely held in the so-called Sher Mandal which Humayun had apparently converted into his library. Legend has it that in his haste to perform obeisance on hearing the cry of the *azan*, Humayun had an accidental fall from the steep staircase of the Sher Mandal. He is said to have died following a temple injury three days later. Humayun's death was as unlucky as anything in his unfortunate life and it had occurred in a 'context which perfectly fitted the man.' Humayun may have done nothing spectacular in India but two of his legacies are extremely relevant to subsequent Mughal history. One came about through his ignoble but fruitful exile in Persia at the pinnacle of its Shia glory under Shah Tamasp. Here, he and his followers not only imbibed literally of all things Persian, but perforce embraced Shiaism, albeit temporarily. The second, and more highly productive legacy that he bequeathed to India was his son Akbar.

Akbar Seizes Power

At the age of thirteen akbar was campaigning in Kalanaur, 300 miles (480 km) away when news of his father's death reached him. Akbar was immediately crowned

Emperor of India and his brilliant career was set on the road by his convincing defeat of Hemu, the Hindu usurper, in the second battle of Panipat. His occupation of Delhi in November of AD 1556 marked the beginning of a rulership of India that in its authority and acumen could be challenged only by that of Ashoka the Great centuries earlier. For Akbar, the city of Delhi seemed to hold no great fascination. No sooner had Hemu's head been despatched to Kabul as proof of Akbar's supremacy, and the threat of Sikander Shah defused by his defeat at Mankot in the Punjab than Akbar decided to make Agra his centre of power. But before power could be really established there were some family affairs to be sorted out. It was largely through the able guardianship of Humayun's General Bairam Khan (now Akbar's regent) that India had been secured for Akbar. Bairam Khan, however, was unable to neutralize the machinations of Maham Anga, Akbar's former chief of nurses. Ultimately, Maham Anga and her son Adham Khan convinced Akbar that he send Bairam Khan on a final Haj to Mecca. During this journey he was murdered by a former Afghan enemy of Bairam Khan. The decks were thus cleared for Maham Anga to promote the interests of her son Adham Khan. The latter, however, by his pettiness and cruelty provoked Akbar into repeatedly throwing him over the Agra Fort walls until he was dead. The mother brought his body back to Delhi. Akbar then allowed Maham Anga to erect a tomb over her son's remains and build a mosque and *madrassa* for herself.

Fig 6.18 Tomb of Adham Khan, near Mehrauli, Delhi, 1561

The Last Octagonal Tomb

For Maham Anga's building efforts in Delhi obviously only local building talent was available. Adham Khan's tomb *(Fig 6.18)* situated south-west of the Qutb complex near modern day Mehrauli is the last redundant effort at further flogging the 'dead house' of octagonal tombs. Though the tomb in its size and execution is undoubtedly as handsome a structure as any built by the Lodi or Sher Shah builders, it shows few signs of creativity. Mercifully, subsequent Mughal builders realized the futility of building any more tombs of this variety. The same is the case with Maham Anga's mosque *(Fig 6.19)* and *madrassa* popularly known as the Khair-ul-Manzil. The mosque is planted directly opposite the eastern gate of the Purana mosque and adjacent to the Sher Shah Gate. The triple arched *liwan* of this mosque

Fig 6.19 Entrance gateway to Maham Anga's mosque, Delhi

takes its cue from the Qila Kunha Masjid of Sher Shah, but in its rather plain and drab plastered exterior, it is unable to reproduce the charm and freshness of the original. However, the composition of the front courtyard and the handsomely proportioned entrance gateway make it a building of some interest.

Persian Resurgence at Delhi

Devoid as Delhi was of direct Mughal patronage, its builders were to reproduce uninspired copies of the old Delhi architecture of the Lodis and Sher Shah. But as soon as Delhi builders were entrusted with the task of erecting a tomb for a true blood Mughal, the undying architectural spirit of Delhi sprang to vibrant life. In building the tomb for Humayun, just outside the envisaged city of Din Panah, Delhi builders once again became the pace setters of architectural fashion in India. The reasons for this sudden resurgence were many. The entire project, apart from having the financial blessings of Akbar, was supervised by Humayun's faithful consort Haji Begum, who had remained by his side right through his travails. These had included a long sojourn in Persia. While in Persia, Humayun had spent a fair amount of his time in sightseeing expeditions that had included the architecture of Safavid Persia. Iranian architecture had obviously made a great impact on the mind of Haji Begum. She now set about the task of translating some of the ideals of Persian architecture to Indian circumstances. The architectural result of the meeting of these two great traditions, the Persian and the Indian, was the erection of one of the more significant buildings of Islamic India. It was destined to exert its influence on the development of Mughal architecture in India over a century later during Shahjahan's reign *(Fig 6.20)*.

Fig 6.20 Red sandstone and white marble tomb of Humayun, Delhi, AD 1565

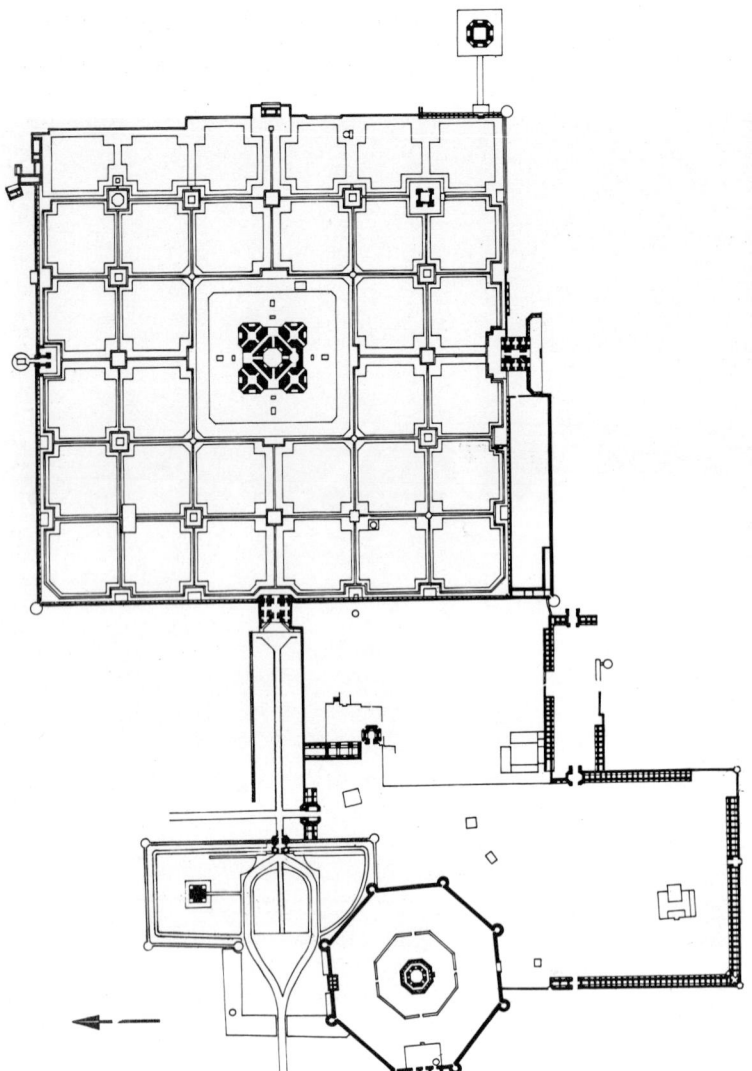

Fig 6.22

Fig 6.21 Typical Japanese gardens

Fig 6.22 Plan of Humayun's tomb built in centre of a Mughal garden, Delhi

Indian, Japanese and Mughal Gardens

The most striking design element of the Mughal tomb is not the building itself but rather the layout of the entire grand complex. The tomb no longer stands alone in the wilderness of a flat and barren piece of land, but is conceived as the heart of a grand symmetrically arranged formal *char-bagh* (or four gardens). The idea of planting the garden around the tomb was a homage by the later Mughals to the nature loving vision of their founding father Babur. Till now, the concept of a garden in India was that of a grove of trees, free flowing rivulets of water and clusters of flowers. The Mughals, instead, organized each of the natural elements within a refined man-made framework of geometrical patterns. Thus, the grove of trees is dispersed into trees planted sentinel-like at strategic points, the rivulets became rigid channels of water laid along the cardinal axes of the building punctuated by fountains at regular intervals, and rows of flowers flagged paths became decorative borders for the grass contained within square quadrangles. The Mughal garden was thus the very antithesis of both

Japanese *(Fig 6.21)* and Indian gardens. While the Japanese garden tended to crystallize the most attractive elements of natural beauty by intense purification of its amorphous forms, the Indian garden was created by merely pruning away the undesirable elements of wild nature to make well shaded groves for human leisure. The Mughal garden attempted to capture natural beauty within a man-made framework highlighting the contrast between the two.

Haji Begum and Humayun's Tomb

To carry out her ambitious task Haji Begum and her retinue of workers, including the chief architect Mirak-Mirza Ghiaz, settled down in a colony close to the selected site on the banks of the Yamuna in the year AD 1556. This settlement has come to be known as the Arab-ki-Sarai even thought it housed craftsmen largely of Persian origin, The Persian craftsmen contributed a great deal of their skills and borrowed an equal amount from local traditions. The plan of the tomb proper *(Fig 6.22)* is not the conventional single chamber but a complex of octagonal halls comprising a central one surrounded by four corner ones, and is inspired as much by Persian models as by classical Hindu *panchratna* planning. In elevation, each of the substantially similar four sides consists of a central rectangular fronton housing a deep semi-arch, flanked on either side by rather squat and cubic masses with chamferred corners. This concept is, undoubtedly, Persian in character. These boldly displayed vast, flat surfaces relieved by deeply shaded arcades and only somewhat softened by the chamferred corners also speak as much of their Persian origin as

Fig 6.23 Front facade of Humayun's tomb, Delhi

does the plain bulbous dome. The white marble dome consciously avoids the traditional and elaborate Hindu *kalasa,* and the apex is marked instead by a simple metal finial. But the idea of raising the edifice over a large and substantial arched stylobate, lifting it off as it were from the gardens all around, the clothing of the monument in beautifully dressed local red sandstone duly contrasted with white recessed marble inlay, the *chajja* and domed canopies over the parapet are positively Indian in origin *(Fig 6.20)*. The platform is, no doubt, inspired by that of Sher Shah's tomb at Sasaram, which in itself follows the tradition of Indian temples. The marble and white sandstone casing is merely a continuation of the Khalji and Tughlaq techniques, while the kiosks were the earliest result of the Hindu-Muslim architectural amalgam. The very substantial and chunky massiveness of the entire structure derives from the solid workmanlike manner in which the tomb has been constructed. This is in striking contrast to the cheaply constructed brick Persian structures that were thereafter covered in the glittering polychrome glazed tiles that almost made the facades like flat painted surfaces. In Humayun's tomb, on the other hand, the richness of the architectural forms and profiles of the arches, the parapets and *minars* is duly highlighted by the complementary taut and rigid bands of inlaid white marble *(Fig 6.23)*. At the same time, the marked absence of the usual *chajjas*, brackets and balconies (except in the kiosks which may well have been latter additions) allow the vast, lightly embroidered surface to stand out with a marked air of freshness and vitality. The whole grand complex thus speaks eloquently of an invigorating merger of two great traditions—the Persian and the Indian. As this merger imbibed more and more of the Indian air, it developed into a highly mature style culminating in the building of the great Taj Mahal within less than a hundred years.

Fig 6.24 Remains of Khan Khanan's tomb at Delhi, AD 1627

The Mughals Shift to Agra

As if to confirm — if further confirmation were required — the necropolis status of Delhi, Akbar sent the body of yet another of his ministers, Atgah Khan, to be buried in Delhi. The builders of Humayun's tomb became busy once again in erecting a diminutive form of their early prototype on a site nearby. Located on the present Mathura Road near East Nizamuddin, this structure is now popularly known as the Khan Khanan's tomb. Bereft as it is of its sandstone casing (which was callously plundered by Nawab Safdar Jung two hundred years later), the building is in a derelict state. Little can be judged of the progress of the art of building in Delhi from this sparse monument. However, its simpler plan form and outer formations appealed more to the subsequent builders of the Taj Mahal, than the rather sprawling mass of Humayun's tomb *(Fig 6.24)*.

To the good or bad fortune of Delhi, however, Akbar did not despatch another body for burial in the royal graveyard, and the former capital city of Islamic India virtually went to seed. According to the British traveller William Finch who visited Delhi just a few decades later, while Humayun's tomb was still 'spread with rich carpets, the tomb itself covered with a pure white sheet of a rich semiane,' on the left stood 'the carcasse of old Dely called the nine castles and fifty-two gates inhabited only by googers.' The Great Mughal, Akbar, had clearly abandoned the city to gypsy settlers and the jackals. The pomp and pageantry had shifted to Agra. It is here that the genesis of the true genius of Akbar as builder both of empire and great architecture was now taking place.

*Entrance gate
to Akbar's
Mausoleum,
Sikandra*

CHAPTER 7

Akbar, the King of Builders

AD 1561–AD 1605

Fig 7.01 Portrait of Akbar

Fig 7.02 Delhi Gate of Agra Fort, AD 1566

Akbar launched his architectural programme in India as his grandfather did by laying out a garden palace seven miles (4 km) south of Agra near a village called Kakrali. There is hardly any trace left of the early buildings of either of the two rulers. While ruins conjectured to be those of Babur's gardens near Dholpur have been excavated recently, no evidence of Akbar's so-called Nagarchain or 'town of peace' exists. This settlement may have been intended to provide temporary relief from the din and bustle of the rebuilding being carried out to make the fortress of the capital city of Agra inhabitable. Akbar had ordered extensive repairs of the ancient Hindu and Lodi fort of Badalgarh on the banks of the Yamuna. Within the Agra Fort from AD 1561 onwards, it is recorded, with the usual exaggeration, that Akbar built 'five hundred buildings of masonry, after the beautiful designs of Bengal and Gujarat, which masterly sculptors and cunning artists of form have fashioned as architectural models.' Of this building activity stretching over a period of about a decade, virtually nothing has survived. Akbar's grandson, Shahjahan, razed most of the buildings to erect his own marble pavilions. What has survived in excellent condition, however, are the ramparts of this fort, its Delhi Gate *(Fig 7.02)*, and a palace inside, all built in local red sandstone. The so-called Jahangiri Mahal was probably built by Akbar for his son Salim at a later stage, and fortunately escaped the eyes of Shajahan's demolition squad.

Akbar's All India Architecture

Even the brief description in the *Ain-i-Akbari* of the numerous non-existent buildings and extant examples are clues to Akbar's architectural vision through his half century of empire and city building. First, in the very employment of 'cunning artists of form' from Bengal, Gujarat and other parts of India. Akbar was launching the scheme of giving a free hand to artists from various parts of the country to exploit their native genius to the fullest. Again, in almost exclusively using locally available red sandstone as building material, Akbar was not only giving vent to his aesthetic partiality but also to his sense of austerity and moderation even in royal buildings. He seems to have been his own Minister of Public Works right through his kingship, and exercised personal control in 'regulating the price of building materials, the wages of craftsmen, and collecting data for framing proper estimates.' What is more, this 'child of a Sunni father and Shia mother, born in Hindustan, the land of Sufism, at the house of a Hindu,' was singularly obsessed with one aspect of his minimal formal education: from the great liberal, Mir Abdul Latif, he had zealously imbibed the principles of 'sulh-i-kul' or religious tolerance. Thus, his builders were rarely commanded to adhere to a style consciously representing a single religion.

Akbar's directives of freedom of expression to his Indian builders were also based on conclusions drawn from what his keen and observant eyes had seen. He had made fairly extensive travels in Hindustan by the time he really got down to building for himself. These had included expeditions to Gwalior, Mandu, Gujarat and large parts of Rajasthan. He had, no doubt, observed the proficiency and the immense diversity of building talent available all over the country. Since, from an early age of his rule, his horizons had envisaged an all-India empire, he had no inclination for promoting any one particular regional style of architecture. Thus, under his liberal patronage, all that was best in the tradition of architecture anywhere in the country was suddenly activated. The only limitation Akbar placed on his architects and engineers was that of economy — a restraint that spurred his artists on to even more sophisticated heights. By this restraint, Akbar the Great Mughal, avoided the customary pitfall of a display of excessive wealth of vulgar extravagance. This is not to say that grand monuments were not produced. Rather, even his more ambitious structures seem to have acquired an air of informal grandeur, rather than monumental oppressiveness.

The Jahangiri Mahal, Agra

The so-called Jahangiri Mahal *(Fig 7.03)* in the Agra Fort is only the precursor to the innumerable structures put up during Akbar's tenure of fifty-six years as Emperor of India. Once again, it is apt to invoke comparison between Akbar's deeds and his grandfather's vision. The only building of India that Babur's critical eye had appreciated, the Man Mandir built at Gwalior seventy-five years earlier, became

Fig 7.03 Jahangiri Mahal in the Red Fort, Agra, AD 1570

Fig 7.04a

Akbar's model for his first major building venture. The Gwalior example probably caught Akbar's fancy for it highlighted the best in the Hindu and Islamic traditions of architecture. The use of coloured glazed tiles and domed canopies on the exterior were as positively of Persian origin *(Fig 7.04a)* as the interior arrangements were Hindu *(Fig 7.04b)*.

The Jahangiri Mahal is essentially a configuration of rooms and chapels freely dispersed around two courts that are aligned along a central axis. The court overlooking the Yamuna river was obviously the private *zenana* court, while near the entrance was the reception court. The two-storeyed front facade of the building, composed of a central, arched opening and deep horizontal *chajjas* over a wall of

*Fig 7.04 (a) Gwalior Fort,
AD 1486–1516, (b) Interior view
of Man Mandir inside the Fort*

Fig 7.04b

Fig 7.05 Facade of Jahangiri Mahal in the Red Fort, Agra

blind arches flanked by octagonal domed turrets, has only a little to do with Man Singh's palace. The interior courts and halls, however, evoke the spirit of the Hindu prototype. The entire scheme consisting of typical Jaina *toranas* sprung across trabeate openings, richly carved stone piers and brackets and inclined struts supporting *chajjas* and roofs *(Fig 7.05)*, is evocative of the palace of Man Singh. Its plan configuration also being similar to that of the Gwalior example is evidence of the fact that the courtly lifestyle of the Mughals and that of the Hindu Rajput chieftains were beginning to approximate each other.

Fig 7.06 Miniature painting showing Akbar instructing masons

Akbar Decides to Build Fatehpur Sikri

Under the circumstances, the Rajput princesses whom Akbar had wedded and brought to his court must have adjusted easily to their new environment. One of them, greatest building enterprise, the new city of Fatehpur Sikri. As is well known, in his desire for an heir, Akbar had visited many Muslim saints to seek their blessings. Ultimately, he found his way to Sikri, situated some 26 miles (41 km) west of Agra. Sikri was a place of some significance to him as his grandfather, Babur, had waged a crucial war in the plains below and built a small mosque here as a 'shukri' to celebrate his victory. Sheikh Salim Chisti, who had established a hermitage on the Sikri hill, predicted the birth of three sons to Akbar. The eldest, Salim, was born to Queen Jodha Bai who had been sent for her confinement to the Chisti household in Sikri. Akbar, who in the 1570s was riding a crest of success with his triumphs in Gujarat and Rajasthan, was convinced by the occurrence of these combined incidents that the Sikri hill would be an auspicious site for a new capital. It is possible that the remains of some existing palaces of the Sikriwal Rajputs — the earlier inhabitants of this area — convinced Akbar of the architectural potential of the site. Akbar, a 28 year-old bold and practical man, set about the planning and building of his city with such determined zest that within less than a decade, the lonely craggy hill of Sikri was transformed into a resplendent new capital city of the Mughals *(Fig 7.07)*. At the zenith of its glory, the city had a population of 2,00,000 and, as Finch records, was larger than contemporary London and Rome. Walking through the numerous palaces, courts and monuments of Sikri, one realizes that nowhere else and at no other time in the history of Indian architecture were buildings of such disparate

Fig 7.07 Fatehpur Sikri, the new capital city of Akbar near Agra

nature so schematically and convincingly knit together into a unified dynamic complex *(Fig 7.08)*. Probably due to the limited space available along the ridge wherein were located the royal palaces, Akbar's architect departed from the conventional idea of building isolated structures linked together by streets. Instead, buildings were conceived as part of a carefully built-up visual and spatial sequence. The spaces enclosed by cloisters and corridors became meticulously designed envelopes for the free-standing structures. One space, one court, one building carefully leads one through the Daulat Khana and the Haram Sarai to the climactic space explosion of the immense courtyard of the Jami Masjid.

Sikri's Enchanting Hour

Various romantic descriptions and equally romantic reasons have been offered for the planning and building of a city that seemed the 'work of a magician's hand,' such as, 'a great complex of palatial, residential, official and religious buildings, so designed and executed as to form one of the most spectacular structural productions in the whole of India;' 'a city containing no streets but an arrangement of broad terraces and stately courtyards *(Fig 7.09)* around which are grouped numerous palaces and pavilions;' and 'some idea may be gained of Fatehpur Sikri during its transient but enchanting hour towards the end of the sixteenth century.' To the romantic, the magic of Sikri was but the 'petrification of a passing mood in Akbar's strange nature, begun and finished at lighting speed while the mood lasted.' But to another more forthright commentator, 'there is little evidence in its layout or composition of any systematic town planning having been put into practice.' The true and technical clue to the planning of Sikri, of course, lies somewhere in between these two extremes. The city is truly a reflection of Akbar's personality and not merely of his despotic whims. There was, behind his apparent waywardness, an extremely shrewd and practical mind at work.

The philosophy that inspired the design and construction of Sikri was only an extension of the spirit that animated Akbar's administration and politics. He realized that his political dream of an all-India empire could not be achieved through sheer

Figs 7.08, 7.09 Fatehpur Sikri, terraces and courtyards surrounded by pavilions and palaces

Fig 7.08

Fig 7.09

physical subjugation of the numerous feudal rulers. Myriads of Rajput and other rulers had to be own through a mixed policy of power of accommodation.

The Politics of Architecture

Fatehpur Sikri, the capital city of such an empire, is but a transparent architectural reflection of its politics. Artisans, architects and builders from all parts of the country were allowed to express their designs freely under the overall control of a sympathetic but powerful guiding hand. The Mughal empire was rapidly built by Akbar through quick marches to various trouble spots, instead of waging ponderously planned wars. In the same way, the city was quickly put together by 'on-the-spot decisions' in many buildings under construction simultaneously at various sites. The location, too, of particular buildings was determined not by a rigid preordained plan but by pragmatic solutions of different problems. Function, orientation, topography, security and aesthetics — each issue was considered, discussed, quickly decided and the work allowed to proceed rapidly. Apparently, the guidelines in each of these considerations were clearly laid down by Akbar himself to avoid unnecessary delays on the part of architects. First and foremost, the interior function of an individual building, whether it was a residence, an office, a workshop, a mosque or a bazaar being easily determined, its plans were accordingly and most forthrightly prepared. Its overall function, too, naturally determined its location in the layout plan, which itself was a flexible document ordaining some broad zoning concept. Service areas, such as the waterworks, serais, and guards' quarters, were located on the outskirts. Public areas like the courts, the Diwan-i-Am, and the Jami Masjid formed a ring around the private audience chambers of the King and Queens' residences, which were located at the very heart, astride the top of the ridge. Orientation was one aspect rigidly adhered to, and so one finds all important structures located along the cardinal axes. While the buildings of a secular nature were installed along the north-south axis, the Jami Masjid was symmetrically erected as required around the east-west axis.

Visual Unity through Sandstone

It would appear that the topography of the undulating rocky ridge was to be least disturbed for practical reasons of economy and viable considerations of environmental control. Thus, the existing natural level of the land chosen for the installation of any building was generally maintained, care being taken only to link it with the adjoining apartments by means of ramps, platforms and staircases. Though the aesthetics of a

Fig 7.10 Sandstone used like concrete and timber panels at Fatehpur Sikri

building project were left largely to the individual group of craftsmen employed in a particular structure, overall visual unity was ensured through various simple strategies. The first and most essential dictate was the pertaining to the use of building material for floors, walls, roofs, lintels, beams, *et al*. The rich, red hue of the stone could be offset only here and there by bands of expensive white marble or at times by blue glazed tiles. The rest of the aesthetics and embellishment was left to the interior decorators of the individual occupants who no doubt 'covered the floors with rich carpets, elaborate silken bolsters, and filled the many alcoves with coloured bottles of perfumes and feminine keepsakes.' What is more, the building technique to be followed was to be the quickest possible. Thus, sandstone, cut to shape in the form of lintels, columns, brackets and roofing tiles, was used much like precast concrete beams and panels or even timber in modern building construction *(Fig 7.10)*. The sandstone building constituents were cut in large stonecutter factories and quickly assembled, often dry, without the aid of mortar. No wonder, then, that the buildings of Akbar's Fatehpur Sikri appear 'to be wooden houses made of stone.'

Symmetry around Multiple Axes

Once Akbar's town planners attuned themselves to his immense desire for speed, the overall aesthetics of urban design, too, were handled unconventionally and with the most fascinating results. Though the general location for a building was determined through the various considerations described above, its exact position

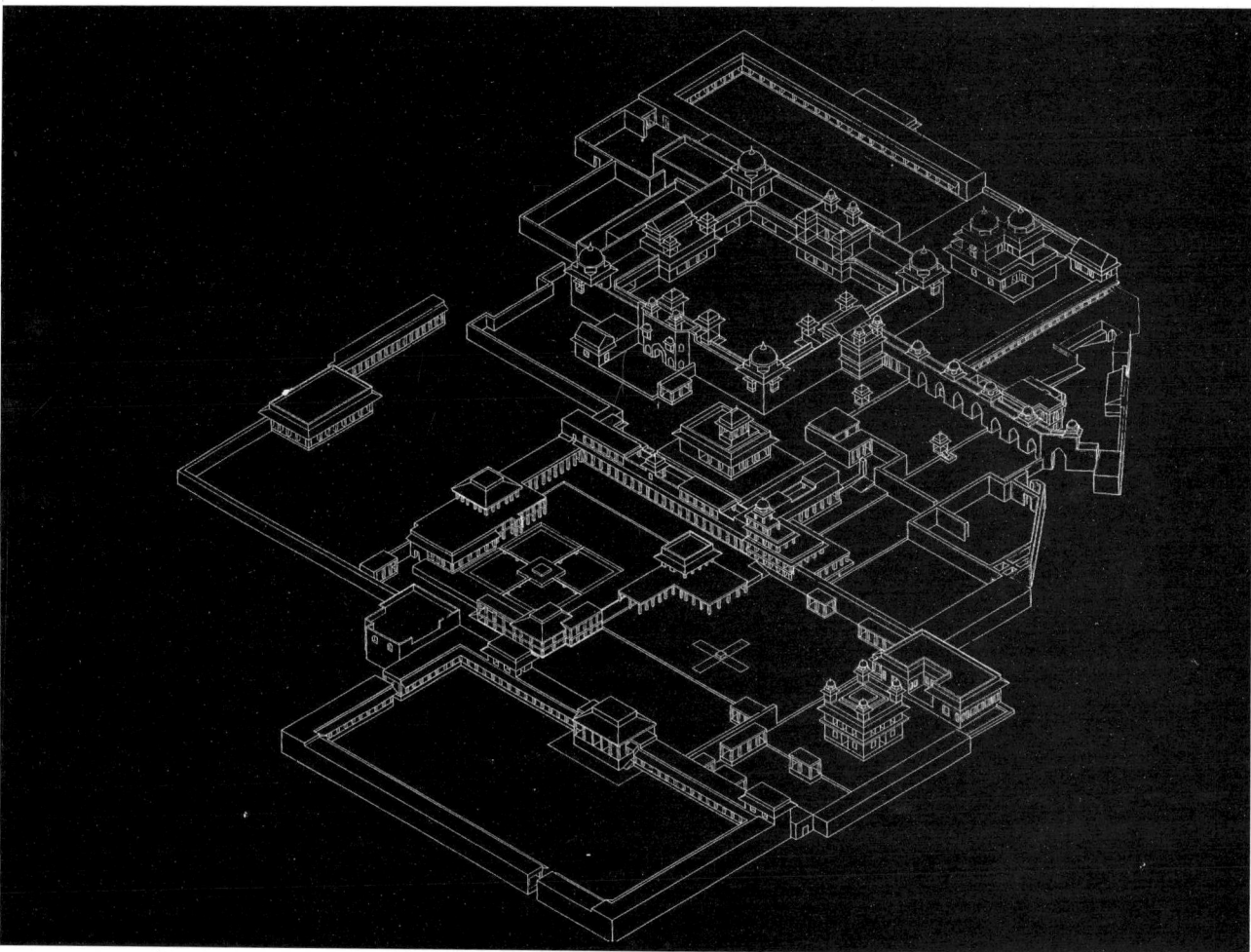

Fig 7.11 Plan of the city of Fatehpur Sikri

was always determined in measurable geometrical relationship to adjoining structures or other prominent features. Thus, even under the seemingly free layout of the city there is an extremely subtle and cunning system of multiple axes at work *(Fig 7.11)*. Every building has a discernible graphic link with another through a series of axes at right angles to each other. Let us study, for instance, the buildings around the Royal Pachisi Complex courts *(Fig 7.11, 7.12)*. Though the Diwan-i-Khas (A) and the Daftar Khana (B) are a perfectly classical and symmetrical composition along the east-west axis, the cubic mass of the Diwan-i-Khas establishes another major relationship with Khwab Gah (C) located at the southern end. In turn, the Khwab Gah symmetrically dominates the Char Chaman Court (D) which in turn is linked along its east-west axis with Maryam's House (E) through a small opening in the court. Again, at the corner of the Char Chaman Court is the exquisite little pavilion of the Turkish Sultana (G), which is the focus of the longitudinal axis of the formal Sultana Garden (H) laid parallel to the outer walls of the Diwan-i-Am. The royal pavilion on the Diwan-i-Am (J) is again linked to this garden along its minor east-west axis.

This process was relentlessly carried right through the royal complex. Just as Akbar followed a dual political policy to run his empire, so his architects, too, adopted the dual policy of containing completely symmetrically designed structures and compositions within a free-flowing composition. It is only for this reason that neither does the free flowing composition fall to amorphous bits, nor do the more grand structures become arid or oppressive symmetrical compositions.

Instant Town Planning

The physical execution of such a theory was probably carried out by first locating the major monuments at the strategic intersections of the main axis. The main axis in turn helped define the courts and platforms built up symmetrically around them. Hereafter, other individual buildings, pavilions, verandas, and colonnades were located according to dictates of need and a subsidiary system of minor axes, as if the buildings were 'movable fittings' adorning the open spaces. It was at any time possible to add an extra pavilion or a hall of prefabricated pillars or make alternations to suit new needs, without in any way detracting from the architectural environment. In fact, these *ad hoc* additions only added to the lively 'stone tent halls' atmosphere of the city *(Fig 7.13)*. Examples of this are the two-storeyed pavilions built around Akbar's Khwab Gah that were obviously improvised in a hurry to facilitate the holding of large audiences attending royal music festivals at the water quadrangle. With an understanding of the above parameters of design and execution, and an idea of Akbar's fascinating personality, a tour of the palaces of Fatehpur Sikri becomes a more enjoyable and intellectually stimulating experience than one of mere amazement at the magic of Sikri. And, what is more, some of the seemingly disturbing curious elements of design of the individual structures fall into understandable patterns.

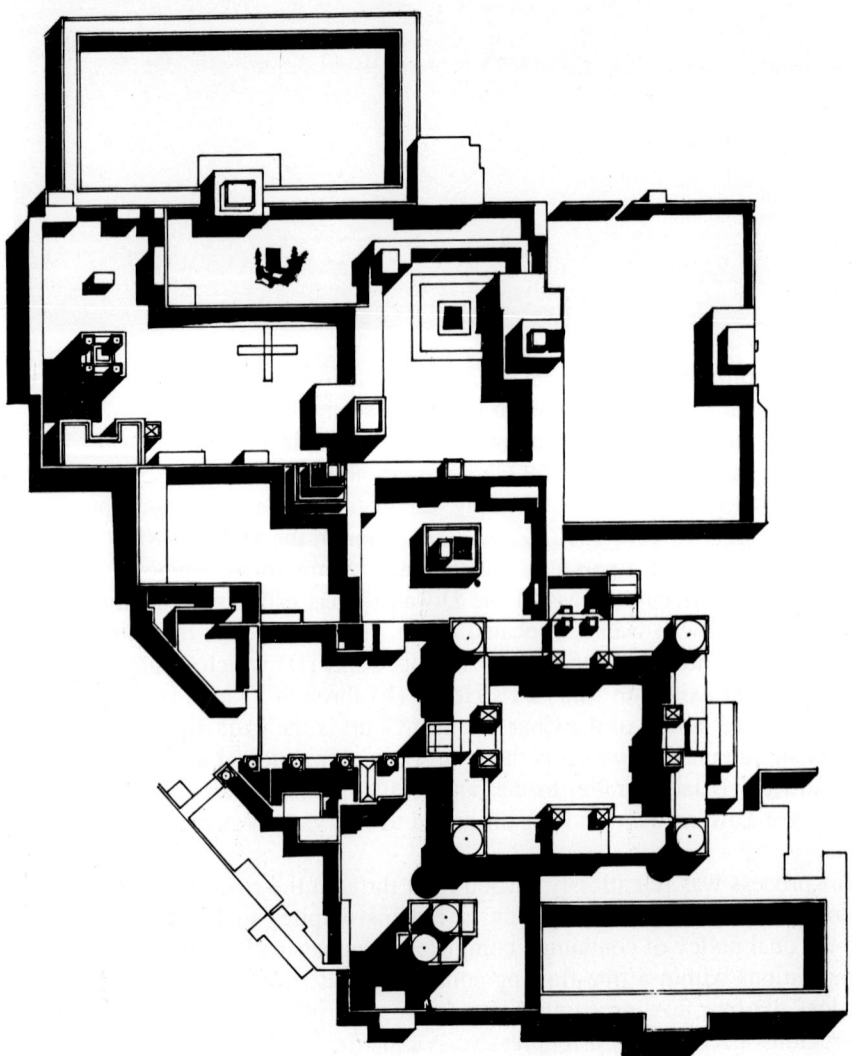

Fig 7.12 Plan of Pachisi court, Fatehpur Sikri

Fig 7.13 Tent-like halls in stone at Fatehpur Sikri

The Diwan-i-Am and Diwan-i-Khas

In the narrow and severely austere oblong of the Diwan-i-Am *(Fig 7.14)*, the location of the public entrances, though breaking the canons of symmetry, becomes immediately meaningful from the point of view of security and function. Neither would latecomers disturb the proceedings underway, nor could a pot-shot be taken at the Emperor through either of the entrance points. This, though, is a simple matter compared to the unique interior of the Diwan-i-Khas. The deceptively simply treated surfaces *(Fig 7.16)* of this unpretentious and modestly planned cubed volume give no clue to its astonishing interior. From the centre of the chamber rises a carved pillar which mushrooms into a gigantic made up of a series of Jaina vaulted brackets that support a circular stone platform *(Fig 7.15)*. From this central platform four aerial bridges radiate along each

Fig 7.14 Diwan-i-Am, Fatehpur Sikri

Fig 7.15 Interior view of the Diwan-i-Khas, Fatehpur Sikri

diagonal of the hall to connect the hanging balconies which communicate with the ground by staircases built within the outer wall thicknesses. 'The Emperor, like a God in the cup of a lotus flower, sat in the centre of the corbelled capital,' while ministers occupied each of the angles. Akbar, who was familiar with Indian mythology, had obviously been inspired by the familiar Buddhist and Hindu imagery of a divine figure seated on a column with a lotus capital — a motif most commonly to be found in the Buddhist caves of Kanheri. Whether this chamber was actually his Diwan-i-Khas or a hall for theological discussions that resulted in the birth of his new religion, the Din-i-Ilahi, is uncertain. However, the entire arrangement is truly a 'poetic realization of a will of power and responsibility.'

That the builders were free to draw their inspiration and revive forms that were almost forgotten in the depths of the rich heritage of the country is evidenced

Fig 7.16 Simple outer surfaces of the cubic volume of the Diwan-i-Khas, Fatehpur Sikri, AD 1570–80

Fig 7.17 Plan and section of Birbal's house, Fatehpur Sikri

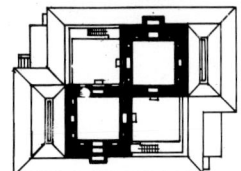

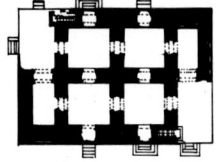

by yet another fanciful structure planted on the western fringe of the great Pachisi court. This so-called Panch Mahal *(Figs 7.18, 721)* is a pavilion of five storeys, the rectangular ground floor being built over eighty-four columns, each different in design. The five storeys above diminish gradually from the northern and western sides, while the eastern and southern facades rise vertically up to culminate in a domed canopy supported over four pillars. Its receding storeys are undoubtedly inspired by the multiple-storeyed Buddhist *viharas*, like the ones that existed in Nalanda and other places of Buddhist pilgrimage. Also called the 'Palace of Winds,' the building was obviously designed for the 'more than 300 wives of Akbar' to enjoy their leisure hours during the long hot months of summer.

Royal Villas of Sikri

Jodha Bai's Mahal, though the largest of the residential structures, is a rather conventional arrangement of a series of palaces and rooms in two floors arranged around a central open courtyard, with the summer and winter palaces located on the northern and southern ends respectively *(Fig 7.20)*. Each of the other smaller palaces displays its own original and extremely practical innovations, almost modern in their basic concept. Thus, the so-called Birbal's House consists of four conjoined rooms with attached porticos on the ground, while only two rooms are built over on the first floor, the flat roofs of the other two becoming open and airy attached terraces *(Fig 7.17)*. Every square niche of its red sandstone surface is covered with Hindu inspired brackets, *chajjas* and intricate carving. Similarly, Maryam Sultana's palace consists of a voluminous rectangular double-height living room *(Fig 7.19)*, at one end of which are built two floors of bedrooms, the upper ones overlooking the double height of the living hall below. In fact, the various residences of Sikri are probably the best examples of very down-to-earth utilitarian and modest royal villas, specifically modulated to suit a refined style of living. Akbar's own bedroom. the Khwab Gah or House of Dreams, contained a large stone platform over which was spread his bed. Lower, along the floors there seem to have been arrangements for watering the room to cool it in summer. The only edifice where Akbar's architects gave way to the conventional style of building was in the great mosque built at the

Fig 7.18 Five-storeyed Panch
Mahal, Fatehpur Sikri

Fig 7.19 View of Maryam's
house, Fatehpur Sikri

Fig 7.20 Courtyard of Jodha
Bai's Palace, Fatehpur Sikri,
AD 1570

Fig 7.21 View of Char Chaman with Panch Mahal behind, Fatehpur Sikri

western extremity of the ridge. It is only here that an almost exclusively Islamic air pervades in the large central courtyard.

The Jami Masjid of Sikri

This massive mosque measuring a fantastic 515 ft × 432 ft (157 m × 132 m) undoubtedly makes it the largest mosque in India. But in attempting a more conventional Islamic configuration, the builders faltered and produced a *liwan* facade that is rather confused and weak. It consists of a central arched fronton which dwarfs the dome behind in the old Tughlaq tradition. Moreover, the central framed arch is completely out of proportion with the rather low-slung side wings. The side wings, too, are formulated out of rather vaguely organized arches of as many as three different spans and heights shaded by a timely, small sloping *chajja*. However, the interior with the central dome decorated on the inside to give an appearance of a stone version of a timber ribbed dome, evokes the spirit of the rest of Sikri *(Fig 7.24)*. The parapets of the cloisters and *liwan* are marked by a row of exquisitely domed *chattris* that held lighted torches at night during festival seasons *(Fig 7.22)*, and add a touch of mirth to the otherwise sombre and uninspired structure.

Akbar himself must have been disappointed with the rather plebian nature of the Jami Masjid of his great city. He availed of every opportunity to impart some liveliness to the drab courtyard and its fortress-like walls. The tomb of one Islam Khan was placed in the north-eastern corner of the courtyard, and subsequently, Salim Chisti too, was buried here in a marble cenotaph built later by Jahangir

Fig 7.22 *Liwan facade of Jama Masjid, Fatehpur Sikri*

Fig 7.23 *Section of Buland Darwaza, Fatehpur Sikri*

(Figs 7.25, 7.27). After seizing the virtually impregnable Ranthambor fort, Akbar decided to commemorate his conquest by adding a massive gateway—the famous Buland Darwaza—to the southern wall of the mosque. The style of gateway indicates that Akbar must have entrusted this task to the craftsmen who had built Humayun's tomb at Delhi. How marvellously they responded to the challenge is written in every stone of the Buland Darwaza.

The Mighty Buland Darwaza

The Buland Darwaza *(Fig 7.23, 7.26)* is not a triumph of mere engineering and structural skills. It is also a unique design solution of a common problem of such ceremonial gateways—that of not merely impressing the viewer with its gigantic size, but at the same time evoking in him sensations of momentary shelter as he passes beneath it. After all, 'men are only six feet high and they do not want portals through which elephants might march.' Thus, the problem lies in meaningfully installing an opening of a modest size within a frame of intentionally immense proportions, and yet maintaining a fluid relationship between the 'crescendo of the great alcove above' and the 'diminuendo of the man height at the base.' The Gothic builders had attempted to solve the problem by splaying their deeply embowed doorways outwards towards the arch of the larger opening. This was a none too elegant nor effective solution. The Roman and Renaissance builders did not even recognize the problem. And so the famous Arc de Triomphe in Paris or the India Gate of New Delhi closer home—though monumental—are not effective gateways.

Fig 7.24 Interior of the grand mosque, Fatehpur Sikri

They do not evoke any feeling of passage and stand as mute and isolated structures. In fact, the Hindu builders of the *gopurams* of the South were rather more successful in piling up a tall mass of pyramidical masonry over a tunnel-like passageway below. However, the intention of these monuments as gateways is apparent only from close view since there is no large portal beckoning the pilgrim from afar. It is only in the Buland Darwaza that the two seemingly contrary visual requirements are satisfactorily reconciled. The huge, almost 50 ft (15.3 m) wide and 100 ft (30.5 m) high arch, resplendent like 'the browed morning,' is backed by a scalloped semi-

Fig 7.25 Facade of Salim Chisti's dargah, Fatehpur Sikri

Fig 7.26

Fig 7.27 Interior of the tomb of Salim Chisti

Fig 7.26 The mighty Buland Darwaza, Fatehpur Sikri (facing page)

domed portal that guides one's vision fluidly down the modest two-storeyed rows of arches and balconies set in sent pentagonal fashion at the base. The central one of these small arches then leads one on to the courtyard of the mosque through a domed passage with attendants' rooms on either side. It is thus that the transition from awesome monumentality to a humble and sheltered passageway, from wide open spaces to the sequestered courtyard is smoothly and satisfactorily concluded.

Sure as the builders were of their ingenuity in relating man to the architectural power of the gateway, they had no hesitation in highlighting its soaring massiveness through every architectonic strategem available. The platform over which the building rises is in itself at an elevation of 42 ft (12.8 m) from the ground below. It is approached by a grand flight of steps that spread out as they descend, thereby providing an appropriately wide base for the lofty precipice-like structure. The central framed arch is flanked by soaring thin minarets and chamferred back broad surfaces that accentuate the verticality of the 134 ft (40.8 m) high portal. Along the top of the gateway is installed the usual merloned parapet and domed kiosks without which no Mughal monument was complete. The rear of the gateway becomes a plain pile of stemmed down masonry. It quietly merges into the cloisters of the courtyard without disturbing the peace and spatial balance of the mosque courtyard. It is ironic and yet true to Akbar's character that having adorned his city with a monument that was 'a miracle of art, imbued with an element of thought, veneration and melancholy that makes up one of those rare sensations of completeness which time cannot impair,' he should have chosen a massive facade to advise its inhabitants that the 'world is a bridge, pass over it, but build no house upon it.'

Akbar deserts Sikri

Eventually, Akbar himself had built too many houses at Sikri. What at one time had been the peaceful hermitage of Sheikh Salim Chisti, had become a busy city. Legend has it that exasperated with the bustle and noise of the new capital, the Saint told the Emperor that either Akbar or he must leave Sikri. Characteristically, Akbar is said to have agreed to leave. But whether this was the reason why Akbar decided to abandon the city, or because his restless spirit could not be tethered too long, or the water supply of the city was drying up, or the collapsed dam in the lake was irreparable, or just that 'his mood had passed' is a matter of speculation. What is conclusive is that by AD 1585, just fifteen years after building the city, Akbar deserted it for ever. Even so, 'the years there had been the richest and the most creative of Akbar's reign. It was here that he established the style of life and culture which would last his family for nearly a century.' Subsequently, a major part of Akbar's life was spent in consolidating the eastern and northern frontiers of his expanding empire. Towards this end, he built a massive fort in Allahabad and set up camp for long periods in the northern city of Lahore. The fort at Allahabad must indeed have been an ambitious effort, for even when Finch visited it in AD 1591, he saw over 5,000 men employed in the building works. Allahabad suffered the fate of many an Indian fortress that was subsequently occupied by the British. A good number of the historic structures were pulled down by the soldiers to make way for hideous rows of barracks. But judging from the one palace of Akbar that somehow escaped the British wrath, the architectural idiom of Sikri had become the royal style. The large, sprawling, square peristylar Allahabad palace is built in the inevitable red stone. Its deep sloping *chajjas*, meticulously crafted brackets and elegant domed canopies all carpentered out of stone, speak volumes for the architecture of Sikri. Though Akbar often made Lahore a base for his expeditions into the north-west and Kashmir, yet nothing of consequence was erected here by him, or his sites were re-built upon by his son Jahangir and grandson Shahjahan. To Akbar, Lahore was essentially a watering station on his way to Kashmir. With its overflowing rivers, lakes and streams, the fabulous valley of Kashmir was becoming a great favourite of the Mughals. Salim, who accompanied Akbar on some of his expeditions from Agra, was particularly enamoured of Kashmir and visited it often later as Emperor Jahangir.

Salim becomes Emperor Jahangir

Before becoming Emperor, Salim, however, was in spasmodic rebellion against his father, Akbar, during the last five years of the latter's reign. There seems to have been good reason on either side for the frequent estrangements. Akbar had at times given signs of his preferences for other possible heirs, and equally had Salim shown calculated disregard for his father's commands. However, Akbar was astute enough during his last brief illness to realize that his preference for anyone other than Salim as his successor would only result in bloody civil war. On the fateful day of 15 October 1605, the Emperor 'motioned (Salim) to wear the royal and turban and to buckle the sword of Humayun,' and with a few last words of advice, breathed his last. Many had entertained apprehensions regarding Salim's ability to rule the huge empire bequeathed to him. Left to himself he would have been content to lead the life of an epicurean gentleman, and this is what he was in fact able to do, without in any way damaging the Mughal Empire. For this he must be thankful to two person. One was the wife he chose for himself a few years after becoming Emperor, the well known Nurjahan who proved an extremely able proxy Empress. And the other, a

Fig 7.28 Miniature painting of Akbar the Great

father who, having turned a mere foothold in the north-west into control of the whole of Hindustan, had also 'effectively changed a military dictatorship to a state administered by an extensive civil service.' The inevitable inertia of the civil service somehow carried through the day-to-day administration, while the necessary monarchial decisions were duly given by the Empress. Jahangir, then, was left with ample time to write his memoirs and follow his own pursuits.

A Renaissance for Painters

Full credit must be given to Jahangir for giving a fresh stimulus to the realm of art, particularly painting. He possessed not only a keen and wide artistic outlook, but also an insatiable passion to build up a visual record of the life around him. Thus, during his rule of about twenty-two years, painters flourished in the same way as

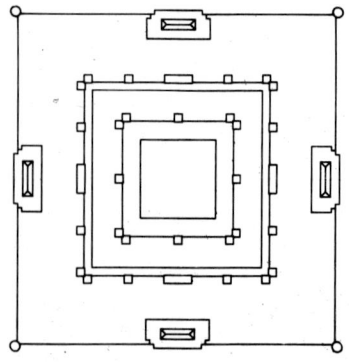

Fig 7.29a

architects and builders had under Akbar. In the process, huge informative albums of paintings depicting not only historic incidents or magnificent personalities but also the flora and fauna of India were assembled. Jahangir's obsession with the two-dimensional art of painting jaundiced his eye for architecture. In fact, during his reign, the erection of buildings was more an excuse for creating flat planes that could then be rendered upon by artists to create the most glorious of surface effects. No wonder then, that to approximate a painter's requirement of starting with a blank white canvas, Jahangir's decorators preferred white marble as the chief building material. Akbar's economic policies, exercised through his astute Marwari finance minister, Raja Todar Mal, had reaped a rich harvest, and hence Jahangir's royal coffers could afford the expenses incurred on painting. But whether the extravagance under Jahangir's control had intrinsic architectural merit is another question. And the question is appropriately posed in the tomb that through a series of historical circumstances ultimately came to be built over the remains of Akbar.

Fig 7.29 Tomb of Akbar at Sikandra near Agra, AD 1612–13, (a) View, (b) Plan

Fig 7.29b

The Enigma of Sikandra

While it is true that Akbar himself had passed the plans of the monument to be built over his grave, in his inimitable way, he had not drawn his inspiration exclusively from the innumerable Muslim tombs around him. This was only natural since all through his living years, he had speculated intensely on matters of religion. At one time, he had more or less abandoned Islam in favour of his own conglomeration of all religions — the *Din-i-Ilahi*. He, therefore, envisioned his tomb as one embodying the ancient Buddhist architectural traditions of India. Experiments in this direction had already been carried out at Sikri in building the Panch Mahal. Now it needed only a symmetrical touch to bring the ideas to more appropriate sober fruition. Thus, Akbar's tomb came to be a 300 square stepped-up formation of masonry, piled up over the grave enshrined in a domed subterranean chamber *(Fig 7.29, 7.30)*. Akbar was thereby giving physical shape to his personal religious heritage of Islam, but under a physical mass that was symbolic of another faith.

That Jahangir could not comprehend Akbar's catholic architecture is not surprising. His approach was neither too orthodox or bigoted, nor too liberal. In good faith he ordered the demolition of much that had been built according to plans passed by Akbar. Although Jahangir may have been genuinely keen to build a tomb for his father 'that should be without parallel in the world.' His good intentions only threw off balance whatever Akbar's architects may have visualized. In fact, the entire monument of Sikandra carries an enigmatic air about it today. Was the tomb intended to be topped by a dome in the classic style? Was the massive base platform with an arched fronton the work of Akbar or Jahangir? Why is the first storey so out of proportion and scale with the basement below? Who laid out the formal garden and the four entrance portals at the cardinal points? Who installed the marble minarets over the entrance gates? Why is the top storey suddenly in white marble while the lower structure is in red sandstone? Some of these questions are answerable and pertinent while some are not.

Fig 7.30 View depicting the stepped-up formation of Akbar's mausoleum, Sikandra

Fig 7.31

Jahangir and a Touch of Marble

The more consciously Islamic elements of Sikandra was certainly Jahangir's. It is quite possible that he enlarged the basement and adorned it with huge arched openings and an arched pylon in the Buland Darwaza style. The entrance gateway, or at least the clothing of it in a rich mosaic of white marble and coloured stones is entirely his handiwork *(Fig 7.31)*. And it is certain that Jahangir spent the bulk of the reported fifteen lakhs of rupees on the penthouse cage of delicately latticed marble planted incongruously over the top. He attempted 'to convert into a lofty structure' what in Akbar's mind had been a playful and virtual picnic pavilion of terraces adorned with domed kiosks and arcades. It is thus that the ambiguity of the intentions of the structure is further highlighted. A visitor to the tomb is unsure whether to accept Akbar's invitation to relax and enjoy or follow Jahangir's dictates to respect and admire. To these conflicting concepts Akbar's grandson later made his own contribution. Shahjahan in his homage to his grandfather planted the four lofty marble minarets on the southern gate. These strikingly resemble those of the famous Taj and certainly belong to this latter school of craftsmen. There is, as we shall see, no evidence that the reign of Jahangir could produce such elegant and mature works of architectural design. Moreover, the location of the minarets over the parapets flanking the main entrance is, to say the least, unusual, and a clear case of fortuitous addition rather than comprehensive design. These minarets were certainly built either as experiments before erecting those at the Taj or immediately thereafter — more probably, the latter.

In spite of not being an altogether convincing architectural monument, Akbar's tomb stands at a crucial crossroads of Mughal history and artistic attitudes. In architectural terms the message is clear. Both Jahangir and Shahjahan were enamoured of white marble as a building material; both were equally indolent and luxurious in their lifestyle, and both had virtually a readymade empire and source of wealth at their command thanks to the ability and vigour of Akbar. No wonder that with the ascent of Jahangir begins the more popularly known and opulent 'reign of marble' of the great Mughals. This reached its climax in Shahjahan's famous Taj Mahal. The long journey from the amorphous curiosity of Sikandra towards the crystalline purity of the Taj is just as visually rich as was Akbar's towards the eclecticism of Fatehpur Sikri.

Fig 7.31 Entrance gateway of tomb at Sikandra (Facing page)

Red Fort,
Delhi

CHAPTER 8

A Century of the Mughals

AD 1605–AD 1707

Jahangir, who ascended the throne in AD 1605, was no great patron of the building arts. His artistic interests lay elsewhere. The almost half century of vigorous and extensive building activity under Akbar was followed by a quarter century of indolent and meagre architectural production under the rule of Jahangir. On the cultural plane Jahangir displayed a keener inclination towards a scientific study of nature than in the creation of great buildings. Nature was his great love and having had it recorded in the form of miniature paintings in vast albums, the closest he came to creating any convincing three-dimensional environment was in the laying out of formal Mughal gardens in his beloved Kashmir.

Gardens in Kashmir

Journeys across the Pir Panjal into this fabled valley had virtually become an annual feature of the pleasure-loving court of Jahangir. In fact, it is said that Jahangir had shifted his capital city to the north-western city of Lahore only because it was closer to the valley of Kashmir, rather than for military or strategic reasons. In the abundantly watered and tree-laden valley of Kashmir, the indefatigable nature loving spirit of Babur sprang once again to life through his great grandson. The environment of the lush valley, too, was far more conducive than the desert-like tracts of Dholpur where Babur had laid out his first garden. Jahangir proceeded to crave out terraced formal gardens at all his favourite spots in Kashmir. These included the foothills surrounding the famous Dal Lake of Srinagar, and also Achabal and Verinag. The results of landscape architecture at each of these places were expressions of the familiar theme of formal Mughal gardens. However, the undulating topography of the land and the abundant and endless supply of water from running mountain streams introduced two new dimensions into the Mughal gardens of the hills. Pools of still waters were replaced by cascades and flowing channels, and the gardens and pavilions were located on gradually ascending levels. Otherwise, the formal geometry of the layout remained the same. There are flagged walkways dividing flat rectangular garden areas hewn out of the foothills into formal square spaces, with chinar trees and flower beds planted in linear patterns along the channelled paths of flowing water. At strategic points of change of level, minimum stone pavilions — sometimes in black stone — were erected. These were perched on short columns over cascades of falling water. Be it Shalimar, Nishat, Achabal or Verinag, the Emperor ordered the same geometric pattern of the Mughal garden — evolved from the square, the rectangle or the octagon — to be superimposed on the natural landscape to create an environment seemingly more comprehensible to human parameters of appreciation *(Figs 8.02a, b)*. But that was the closest Jahangir could come to architecture.

A Coloured Tomb for Jahangir

The few other buildings erected during Jahangir's rule seemed designed more to produce surfaces that could be decorated with mosaics, frescoes and *pietra dura*

Fig 8.01 The Mughal emperor Jahangir

work, rather than enclose exciting spaces or create great architecture. In a gateway to a *sarai* at Jullundur it is apparent that 'the designer has aimed at nicety of detail rather than breadth or strength.' The other more eminent architectural productions that may be attributed to Jahangir's rule, are more the result of the perseverance and devotion of his wife, Nur Mahal Begum (Nurjahan), than Jahangir's own efforts. These are a tomb for her husband in Shahdara (Lahore), and a tomb for her father in Agra, popularly known as the Itmad-ud-Daulah.

A persistently feminine touch is all too apparent in both of the tombs. The one at Shahdara built in AD 1626 is a rather lacklustre 325 ft (99 m) side square structure, at each corner of which are planted 100 ft (30.5 m) high minarets *(Figs 8.03a, b)*. The minarets, though graceful in themselves, are completely out of proportion with the flat central mass. One wonders if even the non-existent marble canopy, which is supposed to have stood in the middle of the square platform, could have added any balance or enlivened the architectonics of this apparently ill-conceived design. The entire structure is constructed of brick with applied colour decoration distributed freely over all its surfaces. To cover such vast brick wall surfaces, 'fresco paintings were used in the interior, inlay coloured work on the pavements and sides of the minarets, coloured glazed tiles on dados in the corridors, and even semi-precious stones enrich the scrolls in the white marble cenotaph.' Yet, in spite of, or rather because of the uninhibited use of so many decorative techniques the end result is more that of a museum of the surface arts rather than a work of architecture. Even the huge 1,500 ft (457.3 m) square garden within which the tomb is set is most conventionally divided into sixteen equal squares with a fountain and pool to mark each intersection. In season, and with each of the parterres adorned with different coloured flowers, the garden may well have been a glorious feast of colour. Without the uncertain aid of blossoming flowers, the vast intended surroundings contribute little to the uninspiring and plain central structure.

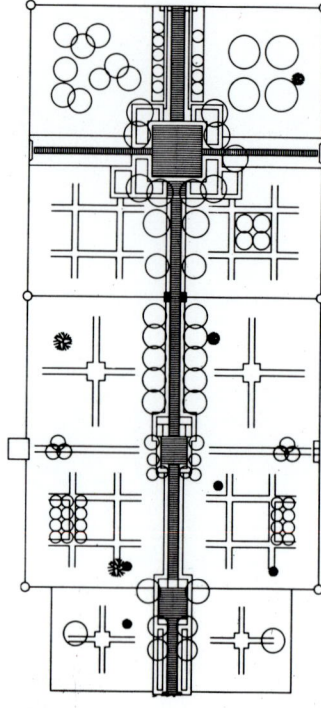

Fig 8.02a

Fig 8.02 A Mughal garden in Kashmir, (a) Plan, (b) View

Fig 8.02b

Fig 8.03a

Fig 8.03 Jahangir's tomb, Lahore, AD 1626, (a) View, (b) Plan

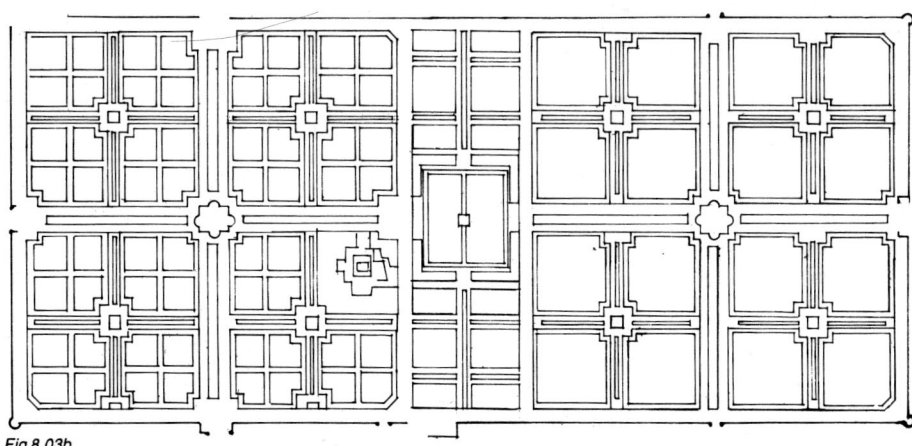

Fig 8.03b

A Marble Tomb for Ghiyas Beg

Nur Mahal's attempts towards building a memorial for her father (who, during Jahangir's time had acquired the title of Itmad-ud-Daulah — 'pillar of the empire') resulted, on the contrary, in an extravaganza of marble with permanent, precious coloured stones replacing the undependable and seasonal flowers of nature. Its design being a miniature version of the one used at Shahdara, this Agra tomb *(Figs 8.04, 8.05)* in its architectural composition, is just as unconvincing as its prototype at Lahore. Fat and ungainly minarets or rather octagonal turrets with cupolas are planted at the corners of an undistinguished 70 ft (21.3 m) side square, single-storeyed structure. Over the roof of the central structure is an equally irrelevant rectangular marble kiosk with *jaalis* all around. Uncertainty of proportion and architectonic intent are the hallmarks of this rather flaccid composition. The weak rectangular shape of the canopy over the middle is as inappropriate to its strategic central location as are the proportions of the so-called minarets. So such so, that even the word minaret — which implies a certain inherent vertical gracefulness — can hardly be applied to these ungainly towers, Obviously, great architecture was not within the capacity of Nur Mahal's builder. Rather, the building was meant to display the accumulated wealth of the royal Mughals through vain feminine channels.

This superficial intention was eminently achieved by the simple expedient of clothing the entire monument in the finest and most expensive white Makrana marble. From a distance, this casket of white marble set amidst gardens and flowing water channels, holds forth the promise of great architectural delights. Unfortunately, Nur Mahal's craftsmen were not quite up to the task of truly fulfilling the promise. Though the adamant marble has been virtually treated like soft muslin and richly 'embroidered' with *pietra dura* work — an art that involved the laying of hard and rare stones such as topaz, onyx, jasper and carnelian into shallow chases cut into the

Fig 8.04 Itmad-ud-Daulah's tomb, Agra

Fig 8.04

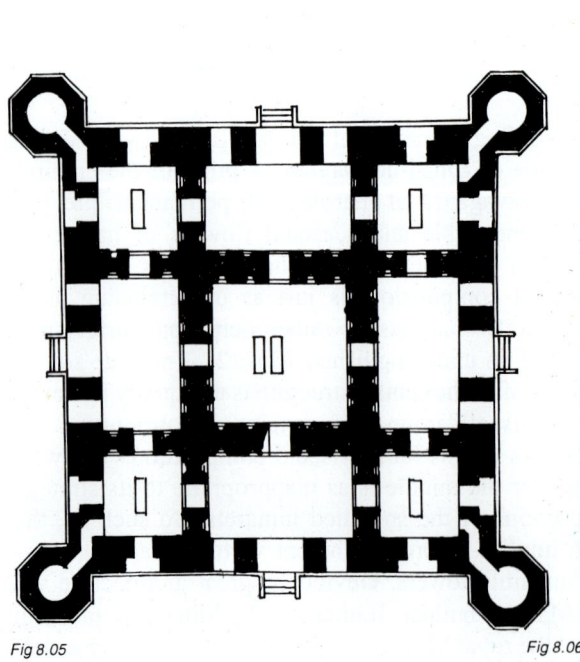

Fig 8.05

Fig 8.06

Fig 8.05 Plan of the tomb of Itmad-ud-Daulah

Fig 8.06 Minaret of Itmad-ud-Daulah's tomb

marble—the workers were so engrossed in the fine and time consuming task of embellishing individual panels that they became oblivious of the very elements of the relationship between applied art and architecture. Thus, floral, geometrical and linear patterns were traced out on the various surfaces as suited the whims of these 'building jewellers.' Whatever architectural spirit there may have been in the admittedly feeble design of the structure was completely lost in the sea of disorganized applied decoration. For example, the changing surfaces of the corner towers are in rapid succession decorated with unrelated bands of vague octagonal patterns at the base, upright and horizontal panels immediately above, followed by realistic pine trees set within rectangular frames, to be finally topped by pseudo geometric floral patterns on the circular drum. The total result is that of a patchwork quilt in which, though each pattern be a masterpiece of craftsmanship, the overall architectural impact is indeed ambiguous.

Emperor Shahjahan

The building in its ambiguity and contradictions is characteristic of the temper of the rule of Jahangir. Though, to some, he was an 'extremely humane and warm personality,' it is also known that he was capable of becoming fiendishly cruel. What is more, during his quarter century of rule, no new frontiers were crossed or positive policies framed except probably that of establishing the practice of shifting court in the summers to the sylvan surroundings of the Kashmir valley. Finally, it was during one of these journeys that Jahangir succumbed to the chronic ailment of asthma, which was undoubtedly aggravated by his equally chronic dependence on opium and wine.

Prince Khurram, his favourite son and successor, was indeed made of sterner stuff than his father. Before he could have himself crowned as Shahjahan (Ruler of the Universe), he had to contend with the wiles of his mother who was more interested in promoting the interests of her own blood relations. In this, Khurram showed his firm hand and dealt mercilessly with his opponents. His own immediate relations, including brothers and nephews, too, were killed, blinded or thrown into jail. Khurram's 'blood stained footsteps to the throne were a precedent which would be quoted to him by his own son, Aurangzeb,' some thirty years later.

Initially, as Shahjahan, he set up court in the Red Fort at Agra. The first six years here were one of peace and tranquillity broken only by the din and bustle of building activity which commenced almost from the day Shahjahan assumed control. During the earliest period of his rule he proceeded to demolish the more austere sandstone structures that Akbar had built within the fort. On the site of each one of these he commenced the erection of halls, palaces and mosques constructed largely from pure white marble obtained from the famous quarries of Makrana, Shahjahan set up such high standards of design, workmanship and detail, that speaking from a purely architectural point of view, his famous 'reign of marble' was indeed a fitting climax to Mughal building activity in India.

The Diwan-i-Am at Agra

One of Shahjahan's earliest actions was the construction of a new Diwan-i-Am *(Fig 8.07)* within the Agra Fort in AD 1627—a structure more befitting the splendour of the new court than Akbar's comparatively modest sandstone buildings. Apparently, Shahjahan's craftsmen in this earliest of their buildings made a radical departure from the Jahangiri style of building—a departure so pregnant with new ideas that its vitality sustained them right through thirty-one years of prolific activity under Shahjahan's rule. Their architecture carried an air of authority and certainty, because Shahjahan's builders grasped the fact that the lustrous white marble obtained from

*Fig 8.07 Shahjahan's Diwan-i-
Am at Agra, AD 1627*

the famous quarries of Makrana 'provides its own decorative appearance owing to
its delicate graining, and any ornamentation requires to be most judiciously and
almost sparingly applied.' Having accepted this intrinsic quality of marble and its
dazzling brilliance, the entire architectural gamut of the period of Akbar and Jahangir
was cast into a far more classical mould under the patronage of Shahjahan. This
involved reducing virtually every building plan to a grid of columns planted at such
spacing that each square could be conveniently spanned by arches springing from
column to column in either direction, to be roofed by a flat ceiling resting over a
continuous peripheral bracket. Thus, even the Diwan-i-Am, in its essentials, is a
large, rectangular, flat roofed hall held up over rows of marble encased columns.

A Method with Marble

The schematic composition having been resolved in this rather straightforward
manner, craftsmen skilled in the meticulous working of marble proceeded to add
lustre and richness to it. And it is in the ornamentation and meticulous detailing of
a marble casing that the craftsmen of Shahjahan, under his watchful eyes, displayed
their superiority over Jahangir's, or rather Nurjahan's indulgent craftsmen. This
excellence was achieved through the strict application of two restraints. One, that
every square inch of the marble was not to be fussed and fretted with; in areas where
a plastic quality had been imparted to this hard material by patient carving and
polishing the material was left to speak for itself. Two, that whatever ornamentation
was applied, even as an intersia of *pietra dura*, this applied decoration was to be
in consonance not only with the architectural intent of the building but also to

highlight its structural quality. Thus, the vertical shaft of a column was not to be covered with irrelevant floral decorations. Rather, minimum black vertical lines could more appropriately highlight the fact that the column was serving the purpose of transferring the load of the building to the ground. In continuation of the same intent, at the two points where this column received the load from the arch and ultimately transferred it to the stylobate or plinth, an appropriate visual appearance of squinched capital and a sort of cushion to mark the points of transition was conveyed by a wider foliated pattern *(Fig 8.08)*.

Thus, the decorative aspect, though rich in craftsmanship, was always subservient to the larger concept of the architectural intent, producing a perfect fusion between applied art and architecture. This fusion became a great architectural treat. Every corner and junction of a building was so meticulously detailed that not even the remotest part of the structure was left to fortuitous design or chance.

The Sensual Classicism of Shahjahan

The purity of detail the marble architecture of Shahjahan could well have been the product of an attitude similar to the contemporary Meisian adage of, 'My God is in my details,' or to the spirit that produced such meticulous and Spartan perfection in ancient Greek temples like the Parthenon. However, the end product of Shahjahan's architecture was more sensuous and rich than either Mies van der Rohe's modern steel and glass boxes, or the austere beauty of the Greek orders. The sensuousness of Shahjahan's buildings is first seen in the breaking up of the stately but austere pointed Tudor arch of Islamic India into the nine cusped arch of the Mughals. Each cusp of this arch follows the other, gracefully resting on the squich-decorated capital of the column. In the dispersion of the arch into such foliations, where in a way, 'each cusp symbolically represented an arch within an arch,' the Mughal builders had adopted the Hindu system of applied sculptural decoration. The complex Hindu system had depended largely on repeating the miniaturized versions of the main architectural form as the theme of the scheme of embellishment. Thus, a flat profile of the temple spire had become the major motif of the decorative scheme of the body of the *shikhara*, as in the temples of Khajuraho. And now, Shahjahan, in an attempt to decoratively enrich his architecture without violating the canons of art, took recourse to the Hindu system of 'ensuring unity in diversity' by repeating the overall forms of architecture, reduced to two-dimensional motifs as the major elements of surface design.

Fig 8.08 A typical Shahjahani marble column in the Agra Fort

Fig 8.09 Bengali thatched roof adapted into the marble medium, Agra Fort

To achieve an equally rich volumetric sensuousness, Shahjahan's builders adopted the typical thatched Bengali roof with its graceful sloping lines into their marble architectural idiom *(Fig 8.09)*. In addition, the rather austere profile of the typical Persian dome of Humayun's tomb was contoured into the more feline grace of the so-called 'onion dome' that the builders of Bijapur had also experimented with. Thus, though extremely classical, Shahjahan's architecture, at the same time, was not lacking in grace and sensuality. In fact, it is this rather rare synthesis of classical and humane values that makes the later period of Mughal architecture in India unparalleled in the world.

An All-India Competition for The Taj

The technique of building vast colonnaded halls to serve as palaces, audience halls and chambers was mastered by Shahjahan's builders in erecting various such structures in the Red Fort at Agra in rapid succession. The more prominent of these are the Khas Mahal or the Diwan-i-Khas, the Shish Mahal and the Musamman Burj in the Agra Fort. The Diwan-i-Khas is similar in plan to the Diwan-i-Am. However, the double columns of this hall are among the most graceful of all those produced during Shahjahan's region. In the design of Musamman Burj, massive square piers instead of columns support the roof, and 'each and every conception of it was executed in the most chaste and exquisite manner.' Through an accident of history that occurred early in his rule Shahjahan posed his architects with probably the most challenging single building commission in the history of the building arts in India. His favourite wife, Mumtaz Mahal, having died during a Deccan campaign at Barhanpur in 1631, Shahjahan determined to build over her grave the most magnificent of tombs ever erected. This, too, of course, was to be clothed entirely in the finest white marble. However, in his bid to get the best talent available in the country Shahjahan floated what was probably India's earliest architectural competition. The end result was the famous Taj Mahal. Now one of the most photographed monuments of the world, its design has been credited by various authorities to various architects. So much, so that even Italian jewellers have staked their claim on it.

Shahjahan's search for a design and an architect may well have been spread all over India, but at the zenith of its glory Islamic architecture in India certainly needed little help from the West. And the Taj, with all its unique qualities is but a fitting, though opulent climax to over four hundred years of activity of tomb building in India. The prototypes of the Taj are not difficult to locate. Shahjahan's designers probably even visited far away Mandu (Hushang Shah's tomb) and Bijapur to furnish the Emperor with complete information on the state of architecture all over India. It is apparent, however, that Humayun's and Khan Khanan's tombs at Delhi received the Emperor's greatest attention. While in its proportions, Khan Khanan's tomb is indeed the most immediate prototype, in its broader aspects it is undoubtedly the tomb of Humayun, built some seventy-five years earlier, that was the first concrete step in the gradual evolution of the design process towards the building of the Taj.

Taj, the Unique

The uniqueness of the Taj lies in some truly remarkable though elementary design innovations carried out by Shahjahan's garden planners and architects. In the true Mughal tradition it was a foregone conclusion that the tomb would be laid out in a garden enclosure. However, for the first time in Islamic India, tomb building was more logically placed at the head of a formal garden rather than in its centre. With this one stroke of genius, the planners added depth and perspective to the first distant view of the monument standing at the end of a marble platformed avenue, water channels and trees *(Fig 8.10)*. In addition, the location of the four free-

*Fig 8.10 The Taj Mahal, Agra,
AD 1634*

standing minarets at the corners of the platform on which the Taj stands added a
hitherto unknown dimension not only to Mughal building but to the very vocabulary
of architectural design. Theoretically speaking, the four minarets provide a kind of
spatial reference frame for the central structure. The function of the minarets is akin
to that of providing a three-dimensional version of the plane rectangle within which
an artist frames his paintings. It is precisely this reference that dissolves what may
have been massive oppressiveness into comprehensible beauty. So much so, that
one does not even notice that the Taj in height is actually more than a modern twenty-
storeyed building. The true design miracle of these minarets, however, can only be
realized when one understands their dual function. On one hand they sequester the
huge monument to impart to it a comprehensible scale; on the other, they prevent its
huge mass from amorphously disintegrating into the horizon. Thus, while both
Humayun's and Khan Khanan's tombs stand 'out somewhat abruptly and isolated
against the surrounding plain.' the Taj, along with its minarets produces an 'outline
that forms a part of rhythmic, undulating total effect.'

A Complete Architectural Experience

It may be redundant indeed to add any more eulogies to the voluminous laudable
writings of historians, architects, literati and poets. What would be more relevant
to note here is the fact that the Taj is amongst the very few works of architecture
that has invited comment and attention from people from all walks of life — from

Fig 8.11a

Fig 8.11b

Fig 8.11 The Taj Mahal, Agra,
(a) Pietra dura work
(b) Roses in marble relief

the intellectual, visionary and poet, right down to the commoner. The revolutionary Urdu poet, Sahir, writing on the Taj Mahal, beseeches his beloved to meet him elsewhere, since to him, Shahjahan had mocked their humble love through such an ostentatious offering to his beloved. Nevertheless, lovers, aesthetes and commoners from near and far continue to throng the Taj. In many ways, the Taj is so complete a work of human architecture that it cannot but fail to arouse an emotional response. The secret lies perhaps in the fact that absolutely nothing has been left to chance. Every aspect of design—be it spatial, architectonic, landscaping, surface decoration or interior design *(Figs 8.13, 8.14)*—has been meticulously detailed and executed.

From a distance, the intriguing skyline of onion domes, canopies and elegant minarets entice and beckon further investigation. An investigation that, at the outset, confronts one with the much recorded and splendid view of the Taj as seen from the entrance gateway, a thousand feet away from the tomb. It would indeed be a soulless person who could turn down the invitation to explore the mysteries of the floating marble vision ahead. And as one walks closer to the monument, across the long marble causeway, one discerns the seemingly plain marble facades of the building alive with an intricate and sumptuous decoration that must be seen at closer quarters. Even the long walk to the monument is enlivened by the changing vignettes and reflections of the Taj in the still waters of the central pool and channel, framed by the upright sentinel-like sombre rows of chinar trees on either side. Having reached the foot of the monument one ascends twenty-two feet (6.7 m) to arrive at the platform over which soar the arched marble facades of the monument.

It is from here onwards that the discovery of the finesse of the Taj begins. The marble casing of the tomb in appropriate panels has been diligently chiselled and polished into delicate relifts of life-life rose bushes bowing their heads gracefully in salute to the Emperor's beloved *(Fig 8.11)*. At the same time, the consummate art of Islamic calligraphy is employed to define and border the soffits of the stately arches in continuous bands, and elsewhere, an intersia of *pietra dura* work in coloured stones embellishes and softens the inherent lustre of the pure white marble. The visitor then either descends into the gloom of the lower crypt containing the graves, or walks to the replicas of the same at an upper level. Here, he is enveloped by an atmosphere as sombre and mysterious as the cave and temple sanctuaries of the Buddhists and Hindus. The journey, a complete experience in architecture, thus seems apparently concluded.

Figs 8.12, 8.13 Surface decoration and detailed designs of the Taj Mahal

Fig 8.12

Fig 8.13

Fig 8.14 *The reflection of the Taj Mahal in pools below*

A Tear on the Face of Eternity

The perceptive viewer immediately realizes, however, that the Taj has much more to offer. To partake of its unending visual delights, he must return to the Taj again and again. The marble in which the monument is encased is 'of such a nature that it takes on incredibly subtle variations of tint and tone' to virtually mirror 'the passing colour of the moment. For every hour of the day and for every atmospheric condition the Taj has its own colour values: from the soft dreaminess of dawn and the dazzling whiteness at midday, to when it is softly illuminated by the brief Indian afterglow to assume the enchanting tint of some pale and lovely rose.' It is on a bright moonlit night, however, that the Taj acquires an almost unearthly quality. Whether seen from across the river Yamuna on the banks of which the Taj stands, or from the southern

gate over the marble paved causeways and water channel,' the hazy glow that emanates from its marble surface, mirrored in the still waters, is a sight no other work of architecture in the entire world is said to equal *(Fig 8.14)*. No wonder then, that poets like Tagore in describing the Taj as a 'tear on the face of eternity,' and photographers using the latest techniques, have been more successful in capturing the magic of the Taj than the architectural historian.

Statistics of the Taj

For the sake of record, the statistical and architectural facts relating to the Taj need to be described. The plan of this whole conception takes the form of a rectangle aligned north and south and measuring an impressive 1900 ft × 1000 ft

Fig 8.15a

(579.2 m × 304.8 m) with the central area divided off into a square garden of 1,000 ft (304.8 m) side *(Fig 8.15)*. The entire composition is enclosed within a high boundary wall with broad octagonal pavilions at each corner. The Taj itself stands on a square 187 ft (57 m) side and 2 ft (6.7 m) high platform. It is flanked on either side by two similar structures: the one on the west is a mosque, and its traditional 'jawaab' on the eastern side served the purpose of a 'mehman khana' or guest house. The mosque and its counterpart of identical design thereby maintain the strict symmetry of the entire composition.

Fig 8.15 Taj Gardens. (a) Plan, (b) View

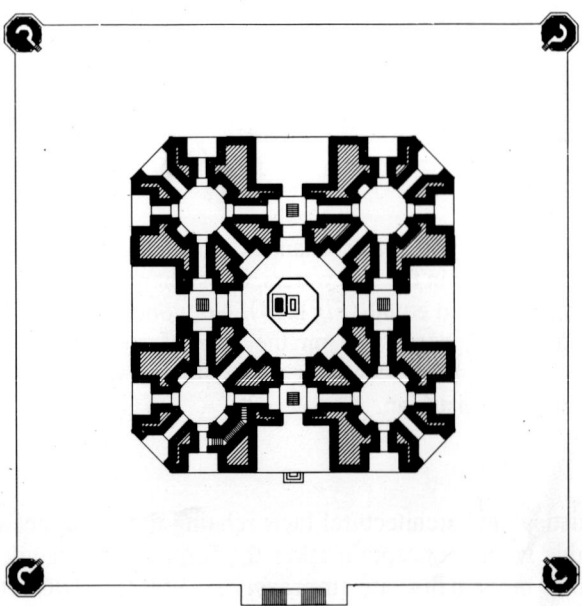

Fig 8.16 Plan of the tomb at the Taj Mahal

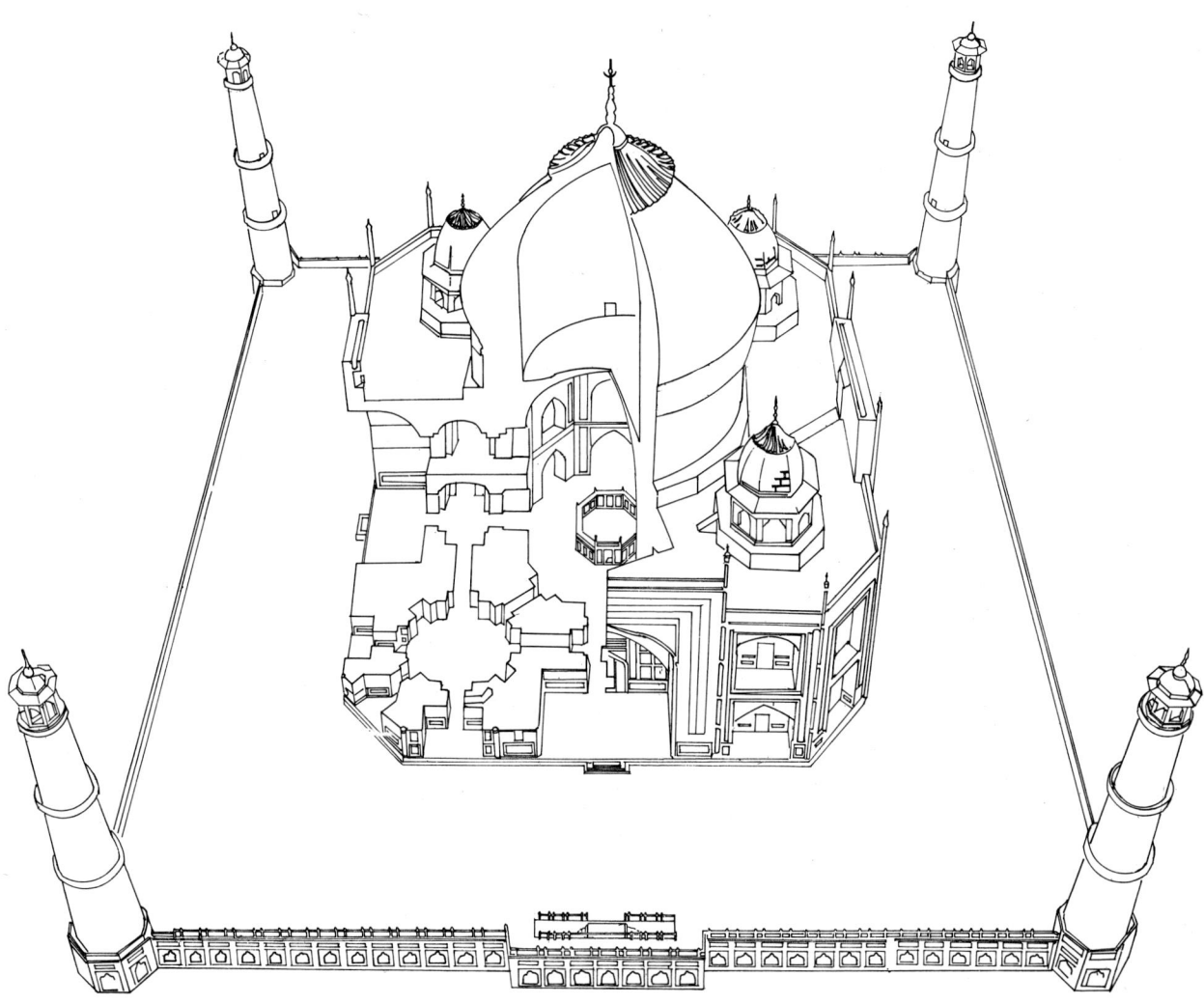

Fig 8.17 Grouping of the Taj domes, Agra

The plan of the tomb *(Fig 8.16)* itself is both a rationalization of the rather loose diagonal arrangement of Humayun's tomb and an enriched though softer version of the plan of Khan Khanan's. Laws of geometry are strictly adhered to in building up the plan, consisting of a central octagonal chamber surrounded by a continuous interlinking corridor, each junction of which is punctuated by a subsidiary octagonal chamber. While the interior with its tunnel and cave-like passage and chambers recaptures the ancient Indian ideals of dark and cavernous spaces, the 187 ft (57 m) high external marble facades of the Taj refreshingly reflect Persian ideals. The designer of the facades judiciously relied on a minimum number of elements. Eschewing all redundant plastic treatment, the architect decided discretely in favour of a completely non-gregarious elevation. In its vast two storeyed treatment, a large central portal arch and a subsidiary smaller arched recess are the only two elements repeatedly and symmetrically woven together into a 'facile grouping, rhythmical disposal and skilful interrelation of each part in the total unity.' The crowning glory of the external visuals is, of course, the shape and volume of the dome. It appears like a 'cloud reclin'd upon his airy throne.' This double dome of the Taj is probably the most refined version of the so-called onion dome already described in referring to the architecture of Bijapur. However, adequately flanked as it is in the Taj — by four domed canopies at each corner — this grouping constitutes probably the most sensual such combination in the entire range of Indo-Islamic architecture *(Fig 8.17).*

Marble Tents of Shahjahanabad

It was symptomatic, however, of the building ambitions of the Emperor that even as this magnificent monument was under construction, Shahjahan in AD 1634 took the decision to shift his capital city to Delhi, a city that provided an essential link between Jahangir's city of Lahore and Akbar's city of Agra-Fatehpuri Sikri. Just four years later, in the time honoured tradition of Delhi—which already and more than half a dozen cities built on its soil—the Emperor proceeded to lay out yet another city—Shahjahanabad. The city was destined to be the last of the great

Fig 8.18 Red Fort built by Shahjahan, Delhi, AD 1639–48

Fig 8.19 Water channels and fountains within the palace of Shahjahan, Delhi

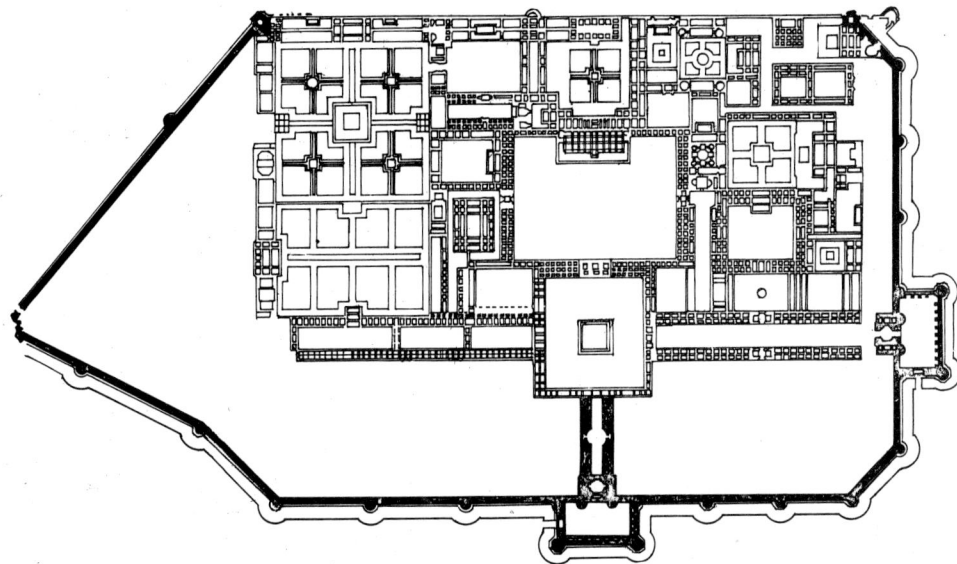

Fig 8.20 Plan of the Shahjahanabad citadel, Delhi

citadels representative of Muslim power in India. In its plan, though, the new citadel fortress at Delhi did not enunciate any new radical directions in town planning *(Figs 8.18, 8.20).* What gives architectural value to the achievement of Shahjahan, however, was the fact that 'it was the conception of one mind executed according to requirements of one authority.' In fact, the plan of Shahjahanabad seems to have been executed in a rather dry and systematic manner. In broad principles, Shahjahan's planners accepted the fundamentals of citadel planning as laid down some two hundred years earlier in the building of Firuz Shah Kotla on a site just a mile down the river. The daring innovations made by Akbar at Fatehpur Sikri went over board after his death. Thus, as would be apparent from a glance, the Shahjahanabad

Fig 8.21 A fountain in the emperor's marble palace, Delhi

citadel consists of a series of garden courts and palaces dispersed on both sides of a central axis composed of the Diwan-i-Am, Diwan-i-Khas and the king's private palace. The whole plan of palaces, courts, hall and gardens is contained within a fortified wall forming an oblong of 3,100 ft (945.1 m) × 1650 ft (503 m) with the military ancillaries located along and just inside the fortified walls.

Almost all structures within the royal Delhi fort were in the form of open pavilions in one storey, their facades of engrailed arches shaded by wide eaves or *chajjas* above which was built a parapet. From each corner of the building arose a graceful domed kiosk. The builders merely prepared various permutations and combinations of these elements to create palaces, halls of audience and hammams and other ancillary chambers. Rightly, the architecture of Shahjahanabad has been referred to as 'marble tent' architecture. The deserted remains of the citadel of Shahjahanabad, as they stand today, are hardly worthy of the glory of the Mughal court. The luxury of life in these 'marble tents' can be realized only when we relate them to the innumerable miniature paintings of the day that depict the palaces furnished, as they were, with all the rich trappings of carpets and canopies, gardens with flowers in bloom, and water cascading and gurgling in the channels and fountains within and around the palaces strung along the riverside *(Figs 8.19, 8.21)*.

It should nevertheless be apparent from other individual structures erected during Shahjahan's time that his builders were more consummate architects than urban designers or planners. Witness, for example the Jami Masjid at Delhi (AD 1644), Wazir Khan's mosque, Lahore (AD 1634), and the Jami Masjid, Agra (AD 1648). Each is an outstanding monument in its own right and manifest of the manifold building activity under Shahjahan's patronage.

Masjids in Agra and Lahore

Though the Lahore mosque takes from the brick and glazed tile buildings of Jahangir's period, it no longer exhibits the architectural uncertainty of the earlier examples. Rather, imbibing the spirit of Delhi, the mosques display the characteristics of a determinate style. In spite of this fact the designers of these brick structures were constrained to design in a manner that would accept the application of decorative flat coloured glazed tiles all over the surface. Nevertheless, the large gateway and four octagonal minarets of Wazir Khan's mosque create the picturesqueness characteristic of other works of Shahjahan's rule. On the other hand, the Jami Masjid at Agra built in honour of the Emperor's daughter Jahan Ara Begum, devoid though it is of the characteristic felicity of Shahjahan's architecture is, nevertheless, a pleasing and strong composition. It is apparent that even before this mosque was under construction, Shahjahan's attention was focused on his new capital. Here, in Delhi, under the Emperor's personal supervision, the construction of the last of the great mosques of Islamic India was under way. This, the Jami Masjid, could truly be considered an appropriate climax to a tradition of mosque building that began with the Quwwat-ul-Islam some five hundred years earlier, and saw the erection of excellent mosques at Mandu, Jaunpur, Ahmedabad and Bijapur in the subsequent period.

Jami Masjid at Delhi

Although no mention is made of my architectural competition or a search for prototypes being conducted prior to the design of the Jami Masjid, it is apparent that a thorough study of mosque architecture in India preceded the finalization of its design. In its detail and sensuousness, it is entirely characteristic of Shahjahan's architecture, while some of its fundamental design ideas are but a revival of old ones. For example, the idea of building the 325 ft (99 m) side platform of the mosque over a raised foundation was an old Tughlaqian device. The application necessitated the installation of gateways

Fig 8.22 Minaret of Jami Masjid, Delhi

Fig 8.23 Intriguing skyline of the Jami Masjid, Delhi, AD 1644–58

in the tradition of the entrance portals to the mosques of Mandu and Jaunpur. The entrance gateway of the Delhi *masjid* leads to an immense red sandstone flagged quadrangle of 325 ft (67 m) side in front of which stands the 220 × 90 ft (67 × 27.4 m) mosque sanctuary. On the other three sides, ranges of cloisters extend their long colonnades, broken by gateways at the cardinal points. It is the graceful onion domes of the sanctuary, the cusped arch fronton and the tall vertically stripped minarets along with the Tudor arched gateways that create the architectural palimpsests familiar to Shahjahan's sumptuous style of architecture *(Fig 8.22)*. True that his mosque lacks the air of gravity of the *masjid* at Mandu, or the militaristic flamboyance of Begumpuri at Delhi, or even the rhythms of the mosque of Atala at Jaunpur. Yet, there is something in its grandeur that is more spectacular than any of its predecessors. It is undoubtedly the typical Shahjahani trait of 'complete lucidity and coherence in its external architectural effect' *(Figs 8.23, 8.24)*.

Fig 8.24 Inner courtyard of the Jami Masjid, Delhi

Aurangzeb Assumes Control

Shahjahan's genius for maintaining an 'external effect' was clearly apparent in the rule and administration. Though externally his was the 'golden period of Mughal rule in India' yet, it is equally apparent 'that there were shadows in the picture' which were well hidden behind the pomp and glory of the court and its equally resplendent architectural productions. In fact, it is the lavish expenditure on the latter which often 'deprived the peasant and artisan of the necessaries of life.' A 'process of national insolvency' had already set in by the time Aurangzeb, the third son of Shahjahan *(Fig 8.25)*, having outmanouvered his brothers, finally imprisoned his father in the Agra Fort. Thereupon, in AD 1658, he assumed control of the empire under the title of Alamgir. Like his father, Aurangzeb lost little time in humiliating or in 'sending out of this world' all possible rivals including Dara Shukoh who had shown a catholicity of religious belief like his great grandfather Akbar. Contrarily, Aurangzeb turned out to be a diehard bigoted Sunni Mussalman. His zeal for Islam and his own strength of character and industrious habits helped him to remain in the saddle for almost half a century — the longest tenure of any Mughal. In fact, by AD 1690, he was 'lord paramount of almost the whole of India, from Kabul to Chittagong and from Kashmir to the Kaveri.' However, his undaunted bravery, grim tenacity of purpose and ceaseless religious bigotry had 'dried up the springs of the tender qualities of the heart.' And without this tender care, all springs of creativity, too, gradually dried up in his empire. No music was heard in his orthodox theocratic court, no poetry was recited and no paintings were made. Good architecture was only yet another victim of Aurangzeb's dry stranglehold over the nation.

The Decline and Death of Mughal Architecture

The rapid decline in the standards of architectural design towards the end of the Mughal era is highlighted by a comparison between the tomb built over the remains of Shahjahan's wife at Agra and the one erected over the grave of Aurangzeb's wife, Rabi Durrani, at Aurangabad, just forty years later. True, that no original production could be expected under the conditions created by Aurangzeb. In fact, the builders' intention was merely to reproduce a smaller replica of the Taj. But even this simple aim could not be achieved. Each and every design element of the Taj, including the minarets and formal gardens was adopted *in toto* in the new design. However, in its compressed proportions and along with some incongruous embellishments such as ornamented-parapets and cusped arches, the entire later production is a mere 'travesty of its immortal prototype *(Fig 8.26)*. It is apparent that the spiritual and human incentive that had made the Taj possible was now just dead wood — as dead and spiritless as the stolid formality of the domes over the Rabi Durrani.

Fig 8.25 Aurangzeb, last of the Mughal emperors

Tomb architecture was not the only form of building to suffer. In spite of Aurangzeb's religious zeal not a single mosque of any consequence was produced. The Moti Masjid, a small chaste mosque, was erected in AD 1622 within the Red Fort, Delhi, so that the Sunni Emperor 'could pray at various times of the day or night without the trouble of a long journey.' Although there is a good deal to commend the excellent workmanship in the finest white marble, the total design, particularly the shape of the domes and entire stilted proportions are symptomatic of the uncertain nature of the times *(Fig. 8.27)*. Almost the same is the case with the only other positive production of this time, the Badshahi mosque at Lahore, built in AD 1674. Though fairly firm and convincing in the stolidity of its brick structured sanctuary which is made up of a central engrailed arched fronton and smaller side arches tapped by three onion domes, the overall effect is rather vague and feeble in the use of attached turrets, minarets and *guldastas*. The battlemented parapet, too, seems rather out of place in a composition that appears to waver between the orthodoxy of Aurangzeb and the picturesqueness of Shahjahan's vision. Also, the details of

Fig 8.26

Fig 8.27

Fig 8.26 Tomb of Rabi Durrani, Aurangabad, AD 1678

Fig 8.27 Moti Masjid built by Aurangzeb, Delhi, AD 1622

decoration on the sanctuary facade 'lack that touch of vitality, that moving play of surface and contrast of light and shade.' It was a sure sign of the 'sap drying up and the architecture becoming stiff and soulless.' No further comment on the state of life under Aurangzeb's rule would be as telling as the recorded fact that 'not a single edifice, finely written manuscript or exquisite picture commemorates Aurangzeb's reign.' Unless one takes note of the mosques that were belligerently erected on the sites of demolished temples in the holy cities of Varanasi and Mathura. Incongruous as their sleek minarets appear in a skyline of full-bodied temple *shikharas*, there is nothing much else to comment on their architecture.

The End of Aurangzeb

Aurangzeb had spent the last twenty-five years of his rule in a military campaign to subdue the Marathas. This adventure virtually bankrupted the royal treasury. In the process both the northern and southern wings of the empire suffered. Soldiers mutinied for pay in the South and chiefs and zemindars in the North openly defied royal authority. Ultimately, the 'Deccan ulcer' ruined Aurangzeb just as the 'Spanish ulcer' had destroyed Napoleon. Aurangzeb himself was apprehensive of the imminent disaster. 'I have not done well to the country and the people, and of the future there is no hope.' But it was too late for him to take any remedial measures. Finally, Aurangzeb died a man broken in body and spirit in March 1707 in Aurangabad, and was buried in an ordinary grave in the courtyard of a religious settlement.

Hereafter, all that remains of Indo-Islamic architecture is the fallout produced by the dispersal of Akbar and Shahjahan's builders and architects to the various provinces of India where they were employed by provincial Hindu, Sikh, and Muslim rulers to build cities and palaces in a form of architecture that, at times, brought out the best in both the Hindu and Muslim styles, and at times resulted in a rather ludicrous mix. Examples of the former are to be found in abundance all over Rajasthan and Central India, and of the latter, in places like Lucknow and even Delhi where Muslim courts, however decadent, continued to exercise feeble rule. However, this gamut of Indian architecture is really a prologue to the beginnings of British and 'modern' architecture in India, a phase of building activity which, by its very importance to modern democratic and secular India, must form part of another volume.

Bibliography

BATLEY, C: *Design Development of Indian Architecture*, DB Taraporevala Sons & Co., Bombay, 1934.

BROWN, P: *Indian Architecture (Islamic Period)*, DB Taraporevala Sons & Co., Bombay, 1942.

BURGESS, J: *Muhammadan Architecture in Gujarat*, Indological Book House, Varanasi, 1971.

CRONESS AND HAYWOODS: *The Gardens of Mughal India, Vikas*, New Delhi, 1973.

COUSENS, H: *Bijapur and its Architectural Remains*, Bharatiya Publishing House, Varanasi, 1976.

EDWARDES, M: *Indian Temples and Palaces*, Paul Hamlyn, London, 1959.

FERGUSSON, J: *History of Indian and Eastern Architecture*, John Murray, London, 1910.

GOETZ, H: *Five Thousand Years of Indian Art*, Methuen, London, 1959.

HAMBLY, G: *Cities of Moghul India*, Vikas, New Delhi, 1977.

HAVELL, EB: *The Ancient and Medieval Architecture of India*, John Murray.

JAIRAZBHOY, RA: *An Outline of Islamic Architecture*, Asia Publishing House, Bombay, 1921.

MAJUMDAR, RC (Ed.): *The Delhi Sultanate*, Bharatiya Vidya Bhawan, Bombay, 1960.

MAJUMDAR, RC, RAYCHAUDHURI, MC, DATTA, K: *Advanced History of India*, Macmillan, London, 1963.

MARSHALL, J: *Annual Reports of the Archaeological Survey of India*, Calcutta, 1903–30.

MICHELL, G (Ed): *Architecture of the Islamic World*, Thames and Hudson, London, 1978.

RAWLINSON, HG: India—*A Short Cultural History*, The Crescent Press, London, 1937.

SPEAR, P: *A History of India*, Vol 2, Penguin Books, Baltimore, 1965.

VOLWAHSEN, A: *Living Architecture: Islamic–Indian*, Macdonald, London, 1970.

Glossary

arabesque: decoration with fanciful interwoven elements

bauli: underground stepped reservoir

bazaar: market

burz (burj): tower, buttress

chajja: an overhang used as a protecting device over wall openings

chhatri: literally, umbrella; small open-sided pavilion-like gazebo

dargah: a Muslim shrine

Diwan-i-Am: Hall of Public Audience

Diwan-i-Khas: Hall of Private Audience

durbar: an assembly around a ruler

guldasta: literally, bouquet of flowers; an ornamental turret

gumbad (gumbaz): dome

hammam: baths

Idgah: Muslim place of prayer on festive occasions

jaali: literally, net; perforated lattice in stone used in Indian buildings

Jami (Jama) Masjid: literally, a mosque for large gatherings; traditionally, the chief mosque of a place

jawab: literally, answer; an identical building installed for the sake of symmetry

kotla: royal citadel

liwan: the pillared and roofed main sanctuary of a mosque

madrassa: school or college building

mahal: palace or royal building

makbara: building over a grave

maqsura: screen of arches of fronting a liwan

masjid: literally, place of prostration; the mosque

mehrab: literally, arch; specifically, the central arch in the western rear wall of an Indian mosque

mimbar: a pulpit in the mosque generally adjacent to the mehrab

qila: fort

qutb: literally, axis or pivot

rauza: a combination of a tomb and mosque in India

sahn: open courtyard of the mosque

sarai: halting place for travellers

Shia: Muslim sect who accept the Prophet's grandson Ali as his true successor

Sunni: Muslim sect who believe in Abu Bakr as the 'elected' successor to the Prophet

wav: stepwell used as summer retreat, mainly in Gujarat

zanana: women's private apartments

Index